Introduction to Theory of Control in Organizations

Systems Evaluation, Prediction, and Decision-Making Series

Series Editor

Yi Lin, PhD

Professor of Systems Science and Economics
School of Economics and Management
Nanjing University of Aeronautics and Astronautics

Grey Game Theory and Its Applications in Economic Decision-Making
Zhigeng Fang, Sifeng Liu, Hongxing Shi, and Yi Lin
ISBN 978-1-4200-8739-0.

Hybrid Rough Sets and Applications in Uncertain Decision-Making
Lirong Jian, Sifeng Liu, and Yi Lin
ISBN 978-1-4200-8748-2

Introduction to Theory of Control in Organizations
Vladimir N. Burkov, Mikhail Goubko, Nikolay Korgin, and Dmitry Novikov
ISBN 978-1-4987-1423-5

Investment and Employment Opportunities in China
Yi Lin and Tao Lixin
ISBN 978-1-4822-5207-1

Irregularities and Prediction of Major Disasters
Yi Lin
ISBN: 978-1-4200-8745-1

Measurement Data Modeling and Parameter Estimation
Zhengming Wang, Dongyun Yi, Xiaojun Duan, Jing Yao, and Defeng Gu
ISBN 978-1-4398-5378-8

Optimization of Regional Industrial Structures and Applications
Yaoguo Dang, Sifeng Liu, and Yuhong Wang
ISBN 978-1-4200-8747-5

Systems Evaluation: Methods, Models, and Applications
Sifeng Liu, Naiming Xie, Chaoqing Yuan, and Zhigeng Fang
ISBN 978-1-4200-8846-5

Systemic Yoyos: Some Impacts of the Second Dimension
Yi Lin
ISBN 978-1-4200-8820-5

Theory and Approaches of Unascertained Group Decision-Making
Jianjun Zhu
ISBN 978-1-4200-8750-5

Theory of Science and Technology Transfer and Applications
Sifeng Liu, Zhigeng Fang, Hongxing Shi, and Benhai Guo
ISBN 978-1-4200-8741-3

Introduction to Theory of Control in Organizations

Vladimir N. Burkov • Mikhail Goubko
Nikolay Korgin • Dmitry Novikov

CRC Press is an imprint of the
Taylor & Francis Group, an **informa** business

CRC Press
Taylor & Francis Group
6000 Broken Sound Parkway NW, Suite 300
Boca Raton, FL 33487-2742

Printed on acid-free paper
Version Date: 20141204

International Standard Book Number-13: 978-1-4987-1423-5 (Hardback)

Library of Congress Cataloging-in-Publication Data

Burkov, V. N. (Vladimir Nikolaevich), 1939-
Introduction to theory of control in organizations / Vladimir Burkov, Mikhail Goubko, Nikolay Korgin, Dmitry Novikov ; edited by Dmitry Novikov.
pages cm
Includes bibliographical references and index.
ISBN 978-1-4987-1423-5 (alk. paper)
1. Management--Mathematical models. 2. Organization--Mathematical models. 3. Control theory--Mathematical models. I. Title.

HD30.25.B867 2015
658.001--dc23 2014037995

Visit the Taylor & Francis Web site at
http://www.taylorandfrancis.com

and the CRC Press Web site at
http://www.crcpress.com

Contents

Introduction

This book explains how methodology of systems analysis and mathematical methods of control theory (including game theory and graph theory) can be applied to organizational management. The theory presented below extends the traditional approach of management science by introducing the optimization and game-theoretical tools for systematic accounting of the special nature of human being as a control object (such as opportunism, selfish behavior, information manipulation). Formal methods are used to construct robust and efficient decision-making procedures (so-called *mechanisms*), which support all aspects and stages of management activity (planning, organization, motivation, and monitoring), over all decision horizons, from operational to strategic management.

We use *mechanism design theory* to construct and study mechanisms of control in organizations. Mechanism design is a branch of game theory. It deals with situations of interaction between a *principal* and a set of *agents*, typically in the presence of asymmetric information. Nowadays, mechanism design underlies modern neo-institutional economics. Recognizing the importance of mechanism design in contemporary economic analysis, in 2007 the Nobel Memorial Prize in Economic Sciences was awarded to Leonid Hurwicz, Eric Maskin, and Roger Myerson, who laid the foundations of mechanism design theory.

Nevertheless, economic theory is only one of many possible applications of mechanism design theory. In addition, mechanism design intersects with the social choice theory and is extensively used in political sciences. Management in all known organizations is based on principal–agent relations, so mechanism design seems to be a natural formalism to study interactions between employees in organizations.

At the same time, formal methods of mechanism design are not too widely spread in *management theory and practice.* One of the reasons is the complexity of underlying mathematics and the lack of wide-scale experimental verification of organizational mechanisms' efficiency in business. Another important problem, which militates against penetration of the methods of mechanism design into management science, is that justification of conflicting interests is just one of the aspects of a management problem. It also includes complex optimization issues and is sufficiently limited by humans' bounded rationality, cost of information acquisition, legal restrictions, and great variety of informal factors.

In this book we introduce an integrated theory, which is based on a version of mechanism design customized to solve the problems of management in organizations. All mathematical models and mechanisms studied hereinafter are motivated by the day-to-day problems met by managers of firms and nonprofit organizations. When possible, we avoid bulk proofs, referring interested readers to the special literature. The focus often goes beyond the stylized framework of a principal–agent game and incorporates the related issues (e.g., optimal production scheduling, performance assessment, and organizational structure design) to construct integrated organizational mechanisms, which solve specific management problems. So, the mathematical and methodological background of the discussed organizational mechanisms is not limited to game theory but also includes systems analysis, control theory, operations research, and discrete mathematics.

This textbook presents an introductory course intended for advanced graduate students specializing in management sciences and in applications of control theory and operations research to business administration. Control theorists will extend their skills with contemporary methods of mathematical modeling of organizational behavior and control, while for business administration students this course will complement their knowledge in organization studies and in the theory of the firm with formal methods improving relevance and efficiency of management decisions by accounting for employees' strategic behavior.

The approach adopted in this textbook is based on the methodology of systems analysis and control theory. This system-based view dictates the structure of the book as it requires an introductory chapter positioning the mechanisms and the model of employees' behavior management on a broader field of control problems in organizational systems of different nature.

The material has the following structure (see Figure 1).

Chapter 1, "Control and Mechanisms in Organizations," gives a general formulation of control problems in

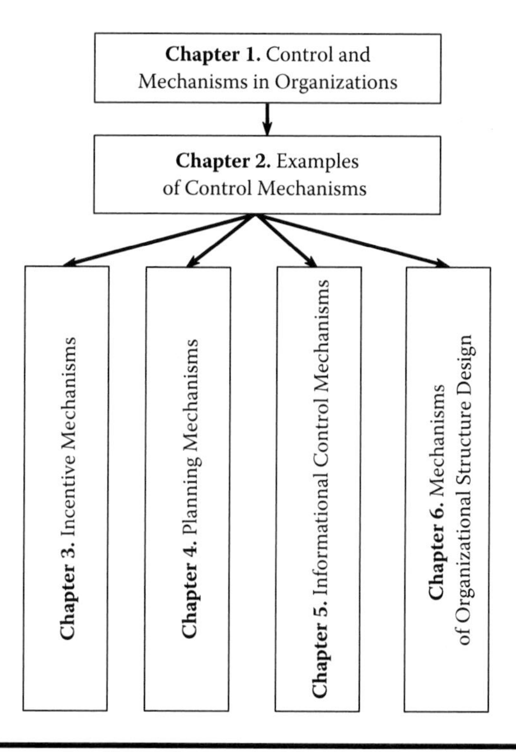

Figure 1 Textbook structure.

organizations, followed by a brief description of the individual and collective decision-making models that underlie models of an organization. The chapter ends with a classification of control problems in organizations.

Chapter 2, "Examples of Control Mechanisms," considers a variety of simple organizational mechanisms to illustrate the main ideas of mechanism design. In addition, their efficiency is evaluated and examples of efficient mechanisms are provided.

In Chapter 3, "Incentive Mechanisms," mechanism design is applied to the problems of providing incentives, that is, the problems of motivating employees to work hard for the success of the organization. As opposed to the traditional agenda of *contract theory*, which tries to overcome the issues originated from incomplete information in agency relations (the models of this sort are postponed to the end of Chapter 4), we limit attention to the more practical problems of group incentives and distributed control, naturally arising in contemporary project-oriented and matrix organizational structures.

Chapter 4, "Planning Mechanisms," deals with the core problem of mechanism design—that of interest coordination when agents have private information. Typical problems of this sort arising in business practice are considered: scarce resource (budgets, funds, and investments) allotment, assigning production plans to production units, management routines where a decision is made on the basis of opinions of a committee of internal experts, and, at the end, incentive problems where employees' reward depends on performance estimates reported by these employees. We explain how the *revelation principle*, the utmost useful implication of mechanism design theory, helps to build efficient mechanisms of management, where agents are motivated to truthfully reveal to the principal the dissipated information about their production capacity, costs structure, and expected returns of investment.

In Chapter 5, "Informational Control Mechanisms," the basic game-theoretical concept of *Nash equilibrium* is extended to incorporate the case when some relevant parameters are not common knowledge among players. The novel concept of *informational equilibrium* is applied to the phenomenon of *informational control*, when the principal manipulates the mutual beliefs of agents to direct them toward common goals.

In Chapter 6, "Mechanisms of Organizational Structure Design," the ideas of organizational studies and of the theory of the firm are incorporated into the mathematical models of organizational structures to explain the origin and the role of complex multilevel hierarchies in business administration. These models provide the basis of optimization techniques of rational organizational structure design, which balance costs of managerial efforts and decentralization costs.

Each chapter includes illustrative examples and ends with exercises on the topic (over a hundred tasks and exercises are provided). At the end, we list several advanced topics for further study, which could be used as essays or student yearly research projects.

Authors' experience accumulated during the years of teaching this course of lectures to different groups of advanced graduate students—some specializing in control theory and mathematical modeling and others focusing on research and development (R&D) management—indicates that for students to take the best advantage of this textbook, the following study courses (see Figure 2) are necessary prerequisites:

1. The group of applied mathematics courses. Single-term courses in system modeling (models of systems, simulation, and system dynamics), decision theory (analysis, linear and nonlinear optimization, probability, and especially basics of game theory [116, 121, 132]), and graph theory [14, 67, 130] (combinatorial analysis and discrete optimization, scheduling and supply chain management) provide the required knowledge of the underlying mathematical framework.

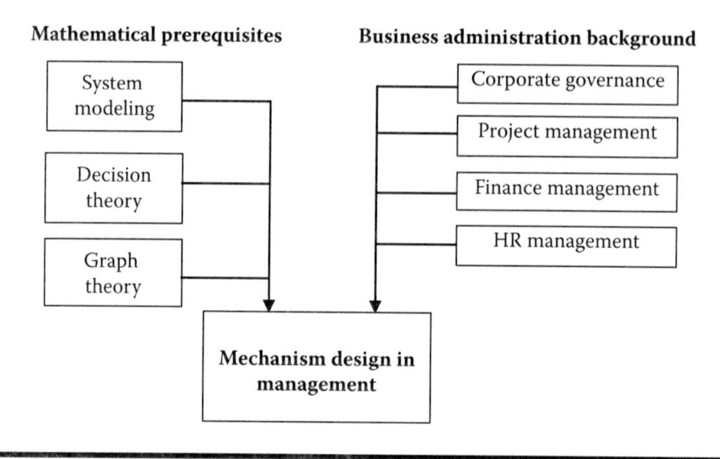

Figure 2 Structure of study courses.

2. The group of courses in different aspects of business administration gives students the necessary background in typical management problems, traditional recipes for their solutions, and also, the best practices [1, 50, 106, 109]. The areas of managerial activity, where accounting for employees' strategic behavior in the course of organizational mechanism design is most important, are corporate governance, project management, human resources (HR) and career management, and corporate finance.

The present book is a complement to the following set of books on the topic of control in organizations:

- *Control Methodology* [125] contains methodological foundations.
- *Theory of Control in Organizations* [127] and *Reflexion and Control* [124] are monographs that may be used as handbooks for detailed theoretical studies.
- *Mechanism Design and Management* [19] explains the unified technology of applied mechanisms design in organizations.

We adhered to the following referencing principle, which seems natural for a textbook: references are given either to the basic monographs (which contain appropriate detailed bibliography; these fundamental classic books allow following historical trends of development) or to the surveys or seminal papers introducing pioneering ideas or certain results employed in the textbook (e.g., theorems and examples).

Prof. A. Chkhartishvili and Dr. S. Mishin participated in the presentation of Chapters 5 and 6, respectively. Authors are deeply grateful to Dr. Alexander Mazurov, who translated this book from Russian to English, for his benevolent and creative efforts.

Chapter 1

Control and Mechanisms in Organizations

Why should we study organizations? With the aim of managing them? With the aim of managing ourselves as members of organizations? We should study them to live within them! Indeed, organizations represent a generic characteristic of humans, exactly as walking on two limbs, having hands, having the ability to speak, having a conscience, and working. Throughout their whole life, humans are absorbed in different types of organizations, from family to global civilization, participating daily in their formation and feeling their favorable or pernicious influences.

Similar to any product of human activity, organizations have a dual nature—that is, a subjective one caused by personal creation and an objective one caused by public creation and predestination. The objective nature of organizations is also defined by the fact that they are organic. Organizations are founded, develop, mature, grow old, and, finally, die. The life of organizations often passes imperceptibly; however, sometimes crises within organizations imply dramatic situations and tragic events for individuals, communities, and generations.

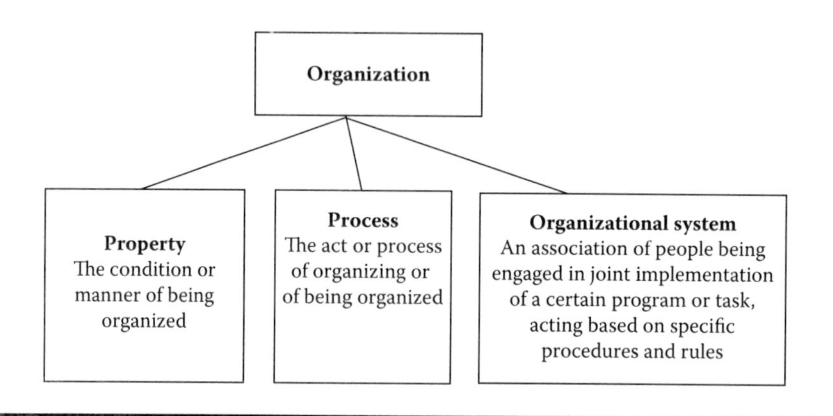

Figure 1.1 Definition of organization.

Let us consider a conventional notion of an organization (see Figure 1.1). According to the definition provided by the Merriam-Webster dictionary, an *organization* is:

1. The condition or manner of being organized
2. The act or process of organizing or of being organized
3. An administrative and functional structure (as a business or a political party); also, the personnel of such a structure

The third meaning of the term *organization* can be extended to the definition of *organizational system* in the following way: "an organization is an association of people engaged in joint implementation of a certain program or task, using specific procedures and rules, in other words, *mechanisms of operation*."

In the sense of organizational systems, a *mechanism of operation* is a set of rules, laws, and procedures regulating interaction between participants in the organizational system. A *control mechanism*, as a set of management decision-making procedures within an organization, appears to be a particular case of a mechanism of operation.

Therefore, the mechanisms of operation and control describe the behavior of the members of an organization*

---

* In this view, control mechanism may be treated as equivalent to control method, since both define *how* control is implemented.

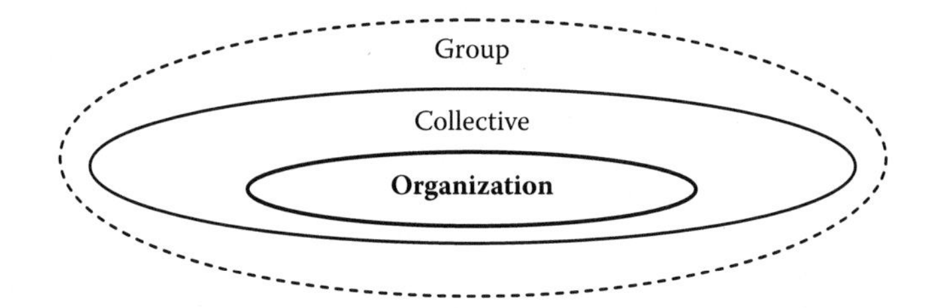

Figure 1.2 Group, collective, and organization.

as well as their decision making. Control mechanisms distinguish an organization from a group (a *group* is a set of people united by, e.g., common preferences, profession, or occupation) and a collective (a *collective* is a group of people performing the same activity). (See Figure 1.2.)

The presence of a definite set of specific control mechanisms in an organization seems appealing, firstly, for the principal because it allows for the prediction of the behavior of the controlled subjects and, secondly, for controlled subjects since it makes the behavior of the principal predictable. The reduction of uncertainty by virtue of control mechanisms is an essential property of any organization as a social institution.

The *principal* can choose a certain decision-making procedure (a certain control mechanism as a relation between his actions, purpose of the organization, and actions of the controlled subjects, i.e., *agents*) only if he is able to predict the behavior of the agents. In other words, he should foresee their reaction to certain control actions. Conducting experiments in real life (using different control actions and studying the reaction of the agents) is inefficient and is almost impossible in practice. We can use *modeling,* which is analyzing control systems based on their models. Being provided with an adequate model, we may analyze the reaction of a controlled system (*analysis* stage) and then choose and apply a control action that ensures the required reaction (*synthesis* stage). The present textbook involves modeling as the basic method of studying organizations.

Organizational Systems as Interdisciplinary: In this textbook, we discuss organizational systems. Hence, a legitimate question arises as to whether other sorts of systems exist and how organizational systems correlate with them. The following interpretation of organizational systems seems possible. Imagine the classification is based on the subject of human activity (nature – society – production). In this case, we may distinguish between:

- Organizational systems (people)
- Ecological systems (nature)
- Social systems (society)
- Economic (technical) systems (production)

Different paired combinations, viz. systems of interdisciplinary nature, emerge at the junction of these classes of systems (Figure 1.3 shows the details):*

- Organization-technical systems
- Socioeconomic systems
- Eco-economic systems
- Normative-value systems
- Noosphere systems
- Socioecological systems

Organizational systems include at least several of the listed elements; therefore, it is reasonable to view them as interdisciplinary ones.

Background: The late 1960s† were remarkable for the rapid development of *cybernetics* and *systems analysis* [5, 16, 105, 133, 154], *operations research* (e.g., see textbooks

* It should be noted that the last three classes have not yet been intensively studied in control theory.

† This time period is remarkable for the simultaneous (and mostly independent) birth of many branches of modern mathematical theories of organizational and economic behavior. We repeatedly refer to this as *time of origin*.

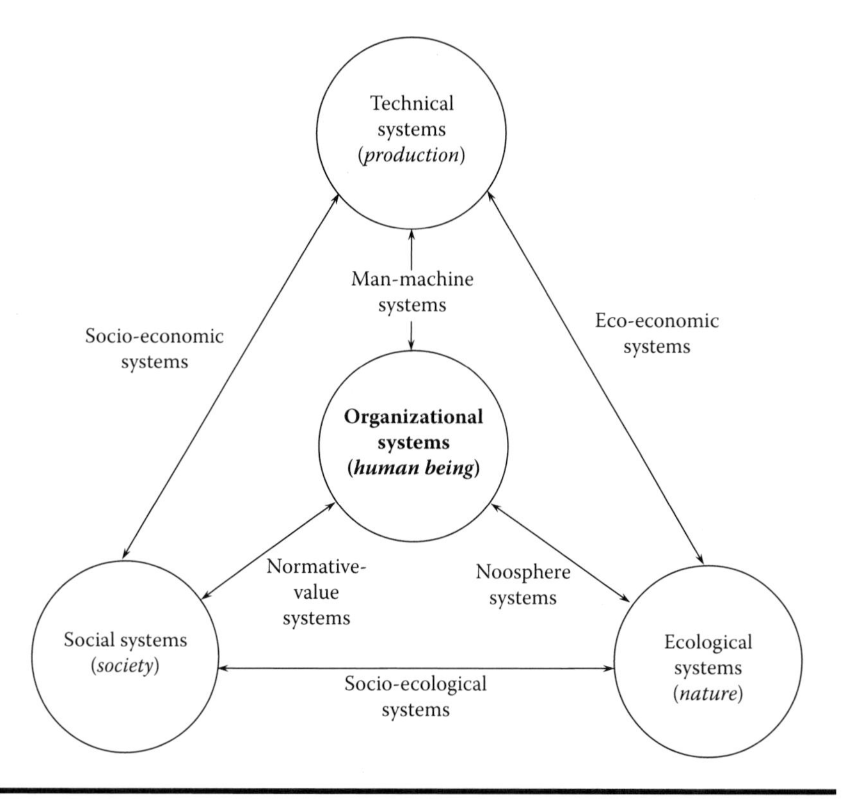

Figure 1.3 Systems of an interdisciplinary nature: Classification.

[147, 153]), and mathematical control theory (automatic *control theory*) as well as for the intensive implementation of their results into technology. At the same time, many scientific research centers endeavored to apply general approaches of control theory to design mathematical models of social and economic systems. Automatic control theory (ACT), *theory of active systems* (TAS) [19, 20, 24, 30–32, 34], *theory of hierarchical games* (THG, [56, 58]), and *mechanism design* (MD [77, 117, 119], including the *theory of contracts* [9, 10, 60, 65, 69], models of *collective decisions* [2, 47, 77, 98], and *implementation theory* [48, 100]) were used (see Figure 1.4). Detailed discussion on the correspondence between the aforementioned (and many other) scientific fields and schools studying organizational control is given in [19].

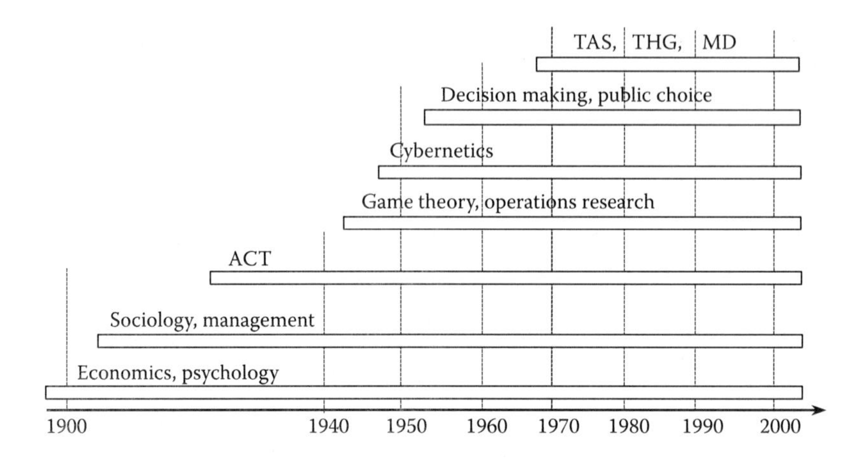

Figure 1.4 Evolution of the concept of an organizational system.

Theory of control in organizations (TCO), which is the subject of the present book, can be understood as an attempt to integrate these theories to develop mathematical and methodological foundations for mechanisms design in organizations [127]. Historically, TCO is the result of further development of the theory of active systems, a branch of control science that developed the methods of accounting systematically for the special nature of humans as a control object (e.g., opportunism, egoistic behavior, and information manipulation). It follows the approach of systems theory and extensively utilizes methods and results of operations research, discrete mathematics, and game theory. Mathematics used by TCO is similar to that used by modern neo-institutional economic theory [99, 103, 107] and *mechanism design* (e.g., game theory [52, 115, 116]), but the range of problems studied by TCO is limited to those of efficient management in different kinds of organizations (e.g., for-profit and nonprofit organizations and governmental structures). For instance, compared with agency theory, TCO is more pragmatic and concentrates primarily on optimal mechanism design rather than efficiency issues. In contrast to management theory and organizational studies [11, 50, 106, 107, 109], TCO is an essentially formal, mathematical theory.

This chapter is introductory; it gives a general formulation of control problems in organizational systems (Section 1.1). In addition, it contains a brief sketch of individual (Section 1.2) and collective decision-making (Section 1.3) models. These models form the basis for the models of behavior by members of organizational systems. Finally, a classification of control problems in organizational systems is proposed (Section 1.4).

1.1 Control Problems in Organizational Systems

Control Activity

Let us start from the psychological foundations of control activity. Following A. Leontiev's theory of human activity [92], consider the basic *structural* (or procedural) *components* of any activity of some individual or collective subject (see Figure 1.5; below *activity* is understood as a purposeful human action) [123]. The chain "a need → a motive → a purpose → an action → a technology → an operation → a result," highlighted by the thick arrows in Figure 1.5, corresponds to a single *cycle* of activity. For convenience, the margins of a subject are marked by the dotted rectangle.

The *need* is defined as a demand or lack of something being essential to sustain vital activity of an organism, an individual, a social group, or society as a whole. Needs of social subjects (i.e., an individual, a social group, and society as a whole) depend on the development level of a given society and on specific social conditions of their activity; these aspects are depicted with the arrow (1) in Figure 1.5.

The needs are stated in concrete terms via *motives* that make a man or a social group act; in fact, a motive is a driver, an object of activity. *Motivation* means a process of persuading, disposing an individual or a social group to fulfill a specific activity or to perform certain actions or operations (see arrow (1) in Figure 1.5). Motives cause formation of a *purpose*

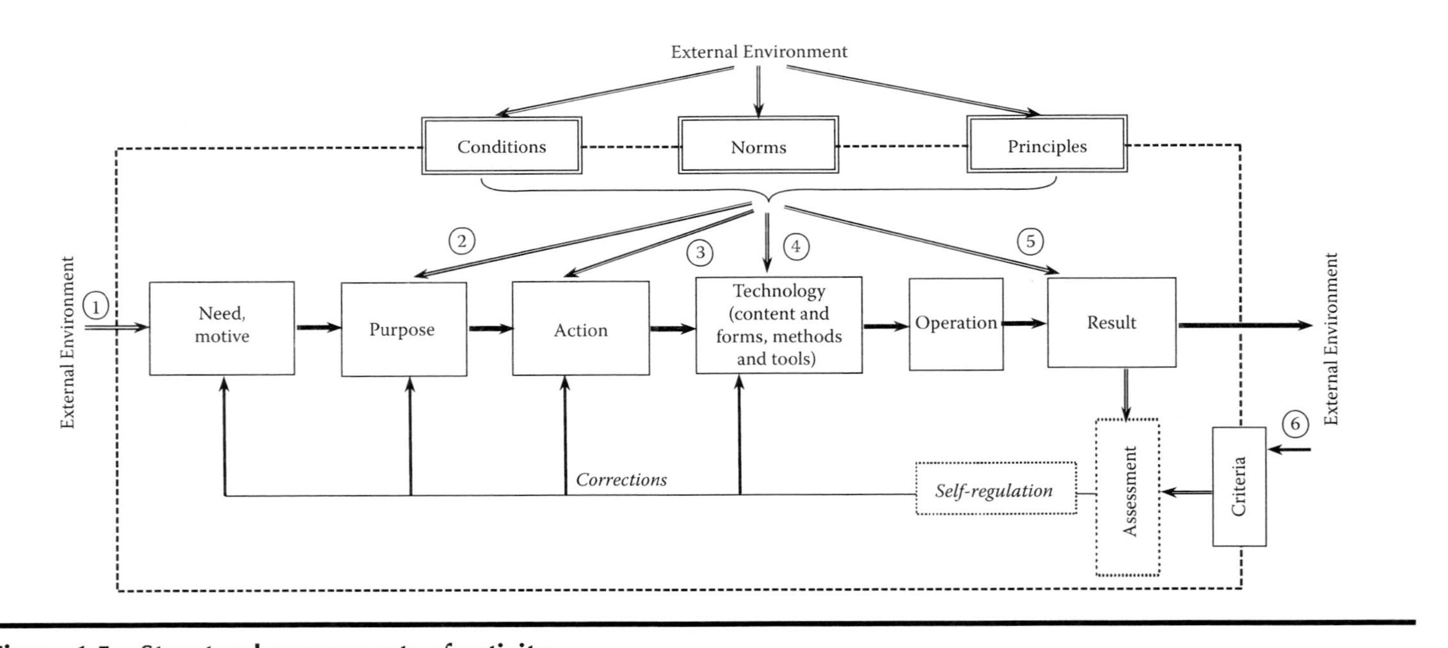

Figure 1.5 Structural components of activity.

as a subjective image of the desired *result* of the expected activity or an action.

The purpose is then decomposed with respect to conditions, norms, and principles of activity into a set of *actions*. Next, taking into account the chosen *technology* (i.e., a system of conditions, forms, methods, and tools to solve tasks), a certain *operation* is performed; note that technology includes *content* and *forms*, *methods* and *tools*. The operation leads (under the influence of an environment) to a certain *result* of the activity. The result of activity is assessed by the subject according to his or her (internal) *criteria* and by external subjects (being a part of an environment) according to their own criteria. A particular position within the structure of activity is occupied by those components referred to as either self-regulation (in the case of an individual subject) or *control* (in the case of a collective subject).

Self-regulation represents a closed control loop. During the process of self-regulation the subject corrects his or her activity with respect to achieved results (see the thin arrow in Figure 1.5).

The notion of an *external environment* (illustrated by Figure 1.5) is an essential category in systems analysis, which considers human activity as a complex system. The external environment is defined as a set of objects/subjects outside the system, whose properties and/or behavior affect the system under consideration or whose properties and/or behavior are affected when the system under consideration changes.

In Figure 1.5 we outline the following factors, which are determined by an external environment [125]:

- *Criteria* used to assess the compliance of a result to a purpose
- *Norms* (e.g., legal, ethical, hygienic) and *principles* of activity, widely adopted within a society or an organization
- *Conditions of activity*
  - Motivational
  - Personnel related

  - Material and technical
  - Methodical
  - Financial
  - Organizational
  - Regulatory and legal
- Informational

Thus, we discussed primary characteristics of activity and introduced the structure of activity. Now we are ready to discuss the notion of control.

Control

To complete the model of a controlled system one has to model also a control system. Control systems are widely different in nature, structure, purposes, and methods of control, resulting in the variety of models of control. So, the main content of many models of organization is, in fact, a sort of control model.

Before discussing the models of control let us give a precise definition of control. We list several common definitions:

Control is the process of checking to make certain that rules or standards are being applied.

Control is the act or activity of looking after and making decisions about something.

Control is an influence on a controlled system with the aim of providing the required behavior of the latter.

There are numerous alternative definitions, which consider control as a certain element, a function, an action, a process, a result, an alternative, and so on.

We would not intend to state another definition; instead, let us merely emphasize that control being implemented by a subject (this is exactly the case for organizational systems) should be considered as an activity. Notably, *methodology* is represented as a science of activity organization, while *control* is viewed as a type of practical activity regarding organization of activity [125]. Such approach, when control is meant as a type of practical activity* (*control activity, management activity*), puts many things into place; in fact, it explains the "versatile character" of control and balances different approaches to this notion.

Let us clarify the latter statement. If control is considered as activity, then implementing this activity turns out to be a function of a control system; moreover, the control process corresponds to the process of activity, a control action corresponds to its result, etc. In other words, within organizational systems (where both the principal and the controlled system are subjects; see Figure 1.7) *control is activity regarding organization of activity* [125].

We can further increase the level of reflexion. On the one hand, in a multilevel control system the activity of a top manager may be considered as activity regarding organization of activities of his or her subordinates; in turn, their activity consists in organization of activity of their subordinates, and so on. On the other hand, numerous business consultants (first of all, *management consults*) are, in fact, experts on organization of management activity.

Consider an elementary† input–output model of a system composed of a control subject (*the principal*) and a controlled

---

* At first glance, interpreting control as a sort of practical activity seems a bit surprising. The reader knows that control is traditionally seen as something "lofty" and very general; however, activity of any manager is organized similarly (satisfies the same general laws) to that of any practitioner, such as a teacher, a doctor, or an engineer. Moreover, sometimes control (or management activity) and organization (as a process, i.e., activity supporting the property of organization) are considered together. Even in this case, methodology as a science of organizing any activity determines general laws of management activity.

† This model is considered elementary as it involves a single principal controlling a single agent. Generalization of the model is possible by increasing the number of principals or agents, adding new hierarchy levels or external parties; for details see Sections 1.3–1.4.

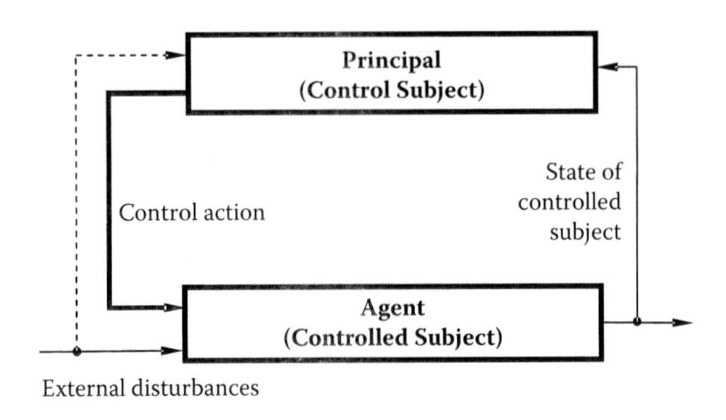

Figure 1.6 The input–output model.

subject* (*the agent*) (see Figure 1.6). Control action and external disturbances form the input of the system, while the action of the controlled subject forms the output of the system. *Feedback* provides the principal with information on the state of the agent.

What elements are chosen by different participants of the system? Let the state of the system be characterized by action of the agent $y \in A$, where A is a set of feasible actions. Suppose the control $u \in U$ belongs to a set of feasible controls U. Consider a criterion of operating efficiency $K(u,y)$ for the whole system; it depends both on the control and state of a controlled subject. In the present textbook we limit attention to scalar preferences and scalar control problems. In other words, we will believe that all functionals map the corresponding sets into the real line: $K(u, y): A \times U \to R^1$ (i.e., multicriterion problems are not considered).

Suppose the response of the controlled subject to a certain control is known. The simplest type of such response is

* Let us clarify the difference between the controlled subject and control object. The term *object* is often used in control theory and decision theory. The object possesses no activity by definition; in contrast, the subject is active and able to make his or her own decisions. Thus, both terms are applicable, but mentioning *subject* we emphasize the humanity of the controlled system. We will use the terms *principal* and *agent* throughout the book, assuming that both are active in the stated sense.

represented by some function of control: $y = G(u)$; here $G(y)$ stands for a model of the controlled subject, describing its response to control action y. As we know $G(y)$, we can substitute it into the criterion of operating efficiency and obtain functional $\Phi(u) = K(u,G(u))$; the latter depends only on control u. The derived functional is referred to as *control efficiency*. Then the problem is to find the *optimal control*, that is, feasible control $u \in U$, which ensures maximal efficiency:

$$\Phi(u) \to \max_{u \in U}.$$

This is an optimal control problem or, briefly, a *control problem.*

A primary input–output structure of the control system illustrated by Figure 1.6 is based on the scheme of activity presented by Figure 1.5; the point is that both the principal and the agent carry out the corresponding activity. Combining the structure of both sorts of activity according to Figure 1.5, one obtains the structure of control activity illustrated by Figure 1.7 [125].

It should be noted that from the agent's point of view the principal is a part of an external environment (numbers of actions in Figs. 1.5 and 1.7 coincide), which exerts an influence for a definite purpose (double arrows (1)–(4) and thick arrow (6) in Figure 1.5; see Figure 1.7). Some components of environment influence may even have a random, nondeterministic character and be beyond the principal's control. Along with actions of the controlled system these actions exert an impact on the outcome (the state) of the controlled system (double arrow (5) in Figure 1.5); see also external disturbances in Figure 1.7. In what follows, the actions under consideration are reflected by *nature uncertainty* or *game uncertainty*.

The structure given by Figure 1.7 may be augmented by adding new hierarchical levels. The principles used to describe control in multilevel systems remain unchanged. However, multilevel systems have specifics distinguishing them from a serial combination of two-level "blocks."

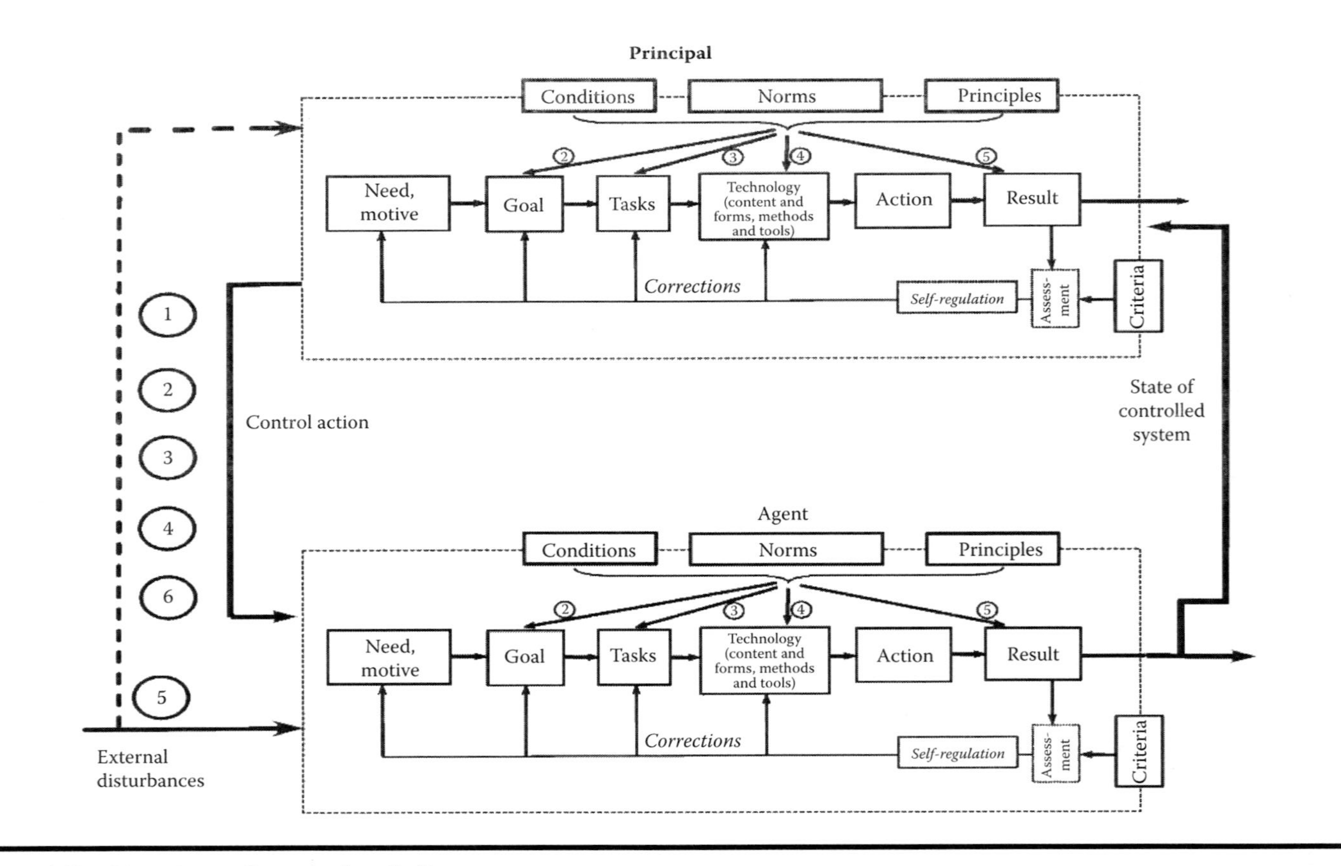

Figure 1.7 Structure of control activity.

Classification of Control

Choosing the proper basis of classification, one may identify the following *classes (methods) of control* for a fixed system [127]:*

- *Institutional control* (administrative, commanding, restricting, enforcing)
- *Motivational control* (motivating the agents to perform necessary actions)
- *Informational control* (persuading, based on revelation of information and formation of beliefs, desires, and motives)†

The following *types of control* (or types of management) may be distinguished based on regularity of controlled processes:

- *Dynamic* control (or *project management*, e.g., change management and innovation management)
- *Static* control (*process management*, i.e., regular and repetitive activity under constant external conditions)

Dynamic control, in turn, can be divided into *reflectory*‡ *(situational) control* and *forward-looking control.* Introducing different bases of classification, one may further extend and detail the list of possible classes and types of control.

The previous classification of types (methods) of control is justified by the following considerations. From the viewpoint of systems analysis, any system can be defined through

---

* The reservation "for a fixed system" is essential here, since possible influence on staff and the structure of the controlled system engenders two additional types of control, viz. staff control and the structure control (discussed hereinafter).

† In some publications on management, the aforementioned types of control are called the methods of organizational, economical, and sociopsychological control, respectively.

‡ Control is called reflectory when a principal reacts to changes or external disturbances as soon as they appear, thus not trying to forecast or influence them. Control is called situational when a certain control action is a priori assigned to each typical situation; every specific situation that happens is classified as some typical one, and the corresponding control action is then implemented.

specifying its components, notably, *staff, the structure,* and *functions.*[*] Therefore, any *organizational system* (OS) is described by specifying:

- *Staff of the OS* (elements or participants of the OS)
- *Structure of the OS* (a set of informational, control, technological, and other relations among the OS participants)
- *Sets of feasible actions* (constraints and norms of activity) imposed on the OS participants; these sets reflect institutional, technological, and other constraints and norms of their joint activity
- *Preferences* (motives, purposes, etc.) of the OS members
- *Information*, i.e., data regarding essential parameters being available to the OS participants at the moment of decision making (choosing the strategies)

The staff determines who is included in the system; the structure describes who interacts with whom, who is subordinate, and so forth. Finally, feasible sets define who can do what, goal functions represent who wants what, and information states who knows what.

Control is interpreted as an influence exerted on an organizational system to ensure its required behavior. Control may affect each of the parameters listed, called *objects of control.*

Hence, using the object of control (this parameter of an OS is modified during a process of control and as the result of control) as a *basis of classification* of control in OS, we obtain the following types of control (see Figure 1.8) [127]:

- Staff control
- Structure control

[*] According to decision theory, any decision-making model includes, at minimum, a set of alternatives available for selection at any given moment, preferences that guide the choice of an alternative by a subject, and information available to this subject.

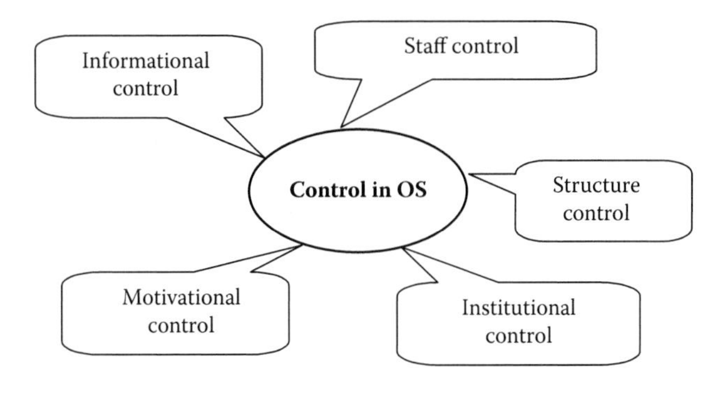

Figure 1.8 Types (methods) of control: a classification.

- Institutional control (control of constraints and norms of activity)
- Motivational control (control of preferences and goals)
- Informational control (control of information available to OS participants at the moment of decision making)

As mentioned earlier, the set of control types (institutional, motivational, and informational control) differs from the previous list only in staff control and structure control. We emphasize that the chosen types of control agree with the adopted diagram of structural components of activity (see Figure 1.5). Indeed, influence of an external environment on the criteria used to assess the activity and also on needs and motives of a controlled subject, forms informational control (see arrows (1) and (6) in Figure 1.5), while influence on the preferences makes up motivational control (see double arrow (2) in Figure 1.5) and influence on the technologies and responsibilities belongs to the sphere of institutional control (see double arrows (3) and (4) in Figure 1.5). Let us briefly discuss specific features of various types of control.* *Staff control* deals with the following issues: who should be a member

* Naturally, in practice it may be difficult to choose explicitly a certain type of control, since some of them could and should be used simultaneously.

of an organization—who should be dismissed or recruited. Typically, staff control also includes problems of personnel training and development.

As a rule, the problem of *structure control* is solved in parallel to that of staff control. Its solution answers several questions, viz. that of control functions to employee distribution, of control and subordination assignment, and so forth.

Institutional control appears to be the most stringent—the principal seeks to achieve his goals by restricting the sets of feasible actions and results of activity of his subordinates. Such restriction may be implemented via explicit or implicit influence (e.g., legal acts, directives, and orders) or mental and ethical norms, corporate culture, and so on.

Motivational control seems "softer" than institutional control and consists of purposeful modification of preferences of subordinates. This modification is implemented via a certain system of penalties and/or incentives that stimulate a choice of a specific action and/or attaining a definite result of activity.

Compared with institutional and motivational control, *informational control* appears the softest (most indirect) type.

Forms of Control

Choosing various bases of classification, one may define different forms of control. Depending on the structure of control systems, we separate out:

- *Hierarchical control* (control system possesses a hierarchical structure, provided every subordinate has a single superior)
- *Distributed control* (the same subordinate may have several superiors, e.g., matrix-type control structures)
- *Networked control* (distribution of control functions among elements of a system may vary in time)

Depending on the number of controlled subjects, we distinguish

- *Individual control* (control of a single agent)
- *Group control* (control of a group of agents, based on the results of their joint activity)

Depending on whether control is adjusted to individual characteristics of the controlled subject, one can separate out:

- *Unified control* (when the same control mechanism is applied to a group of subjects, heterogeneous, in general)
- *Personalized control* (when control depends on individual features of each controlled subject)

Management (control) tools (e.g., orders, directives, instructions, plans, norms, and regulations) are not considered in this textbook. Their detailed description may be found in any textbook on management.

Management (Control) Functions

Following Fayol, let us single out four *primary functions* of control, notably planning, organizing, motivating, and controlling. Continuous sequence of implementing these functions makes a management activity cycle (see Figure 1.9).

The stated management functions are common for process- and project-based management and correspond to the structural components of activity. Let us elucidate this.

Process-based management includes the following primary functions: *planning, organizing, motivating* (stimulating), and *controlling.*

Project-based management has the following stages of a project life cycle:

- *Initiation* (concept), that is, data collection and current state analysis; goals setting, tasks decomposition, criteria selection, requirements and constraints (internal and external), definition, assessment and approval of a project framework

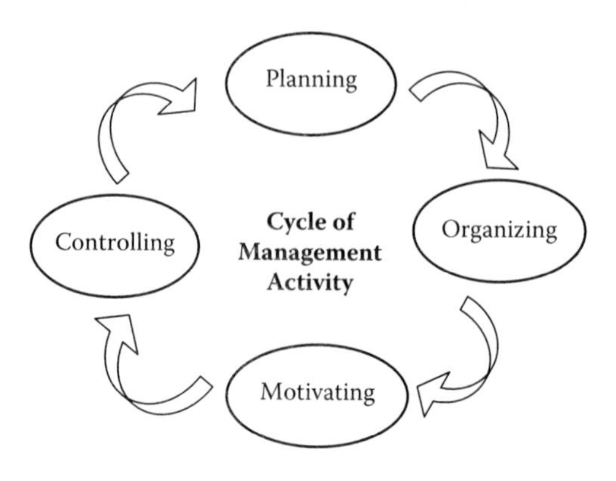

Figure 1.9 Management activity cycle.

- *Planning and design*, that is, team building and training, project charter and scope development, strategic planning, holding tenders, contracting and subcontracting, project draft design development and approval
- *Execution*, that is, starting up a project management system developed at the previous phase, organization and coordination of project activities, starting up an incentive system for project team members, operating planning, procurement, and operating management
- *Monitoring and controlling*, that is, measuring ongoing project activities, monitoring project performance indicators against a project performance baseline and a project plan, performing corrective actions to address issues and risks, change management (adjusting project plans and budgets)
- *Completion*, that is, planning a process of project completion, checking and testing project deliverables, customer staff training, formal acceptance of the project, project performance assessment, project closing, and dissolving the project team

There is an obvious correspondence between the first four project phases and four management functions of planning, organizing, motivating, and controlling.

Table 1.1 Control Types and Functions

Control Types	*Control Functions*			
Process-based management	planning	organizing	motivating	controlling
Project-based management	initiation	planning and design	execution	monitoring
Controlled components of activity	purposes	technology	motives	results

Finally, we can identify yet another correspondence between structural components of activity explained in Figure 1.5 (*a motive, a purpose, technology,* and *result*) and four basic management functions (see Table 1.1).

Hence, one may speak about the following general control functions: *planning, organizing, motivating,* and *controlling.*

Thus, we have formulated the control problem in its general setting. To realize how this problem is posed and solved in every specific case, let us consider a general technology of control in organizational systems.

A Technology of Solving Control Problems in Organizational Systems

We understand *technology* as a set of methods, operations, and procedures that are successively applied to solve a specific problem. Note that technology of solving control problems discussed herein covers all stages, from OS model construction to performance assessment during adoption (see Figure 1.10; for better clarity stages' back couplings are omitted) [125].

The *first stage* (model construction) consists of a description of a real OS in formal terms—specification of staff and the structure of OS, goal functions and sets of feasible strategies of principals and agents, their information, and, for example, the

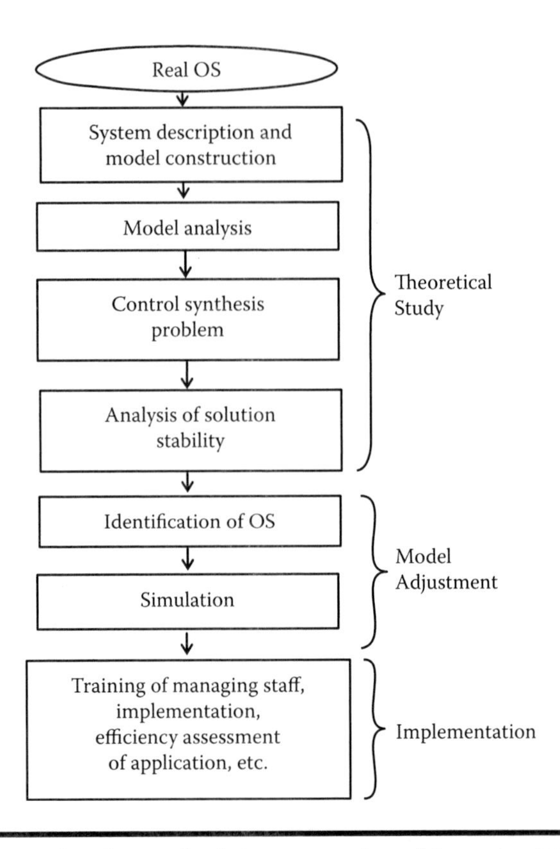

Figure 1.10 Technology of solving control problems in OS.

order of operation and behavioral principles. This stage substantially involves the apparatus of game theory; as a rule, the model is formulated in terms of the latter (see Section 1.3).

The *second stage* (model analysis) lies in studying the behavior of principals and agents under certain control mechanisms. To solve a problem of game-theoretic analysis means to find equilibria of the game of the agents given a fixed control mechanism, chosen by the principals.

After solving the analysis problem (i.e., being aware of behavior of agents under various control actions of principals) one may proceed to the *third stage*. First, one solves the *direct control problem*, that is, the problem of optimal control actions synthesis (find a feasible control ensuring the maximal efficiency). Second, one solves the *inverse control problem* (finding a set of feasible controls rendering the OS to the desired state). Control efficiency criterion is represented by the maximum (when agents can be considered benevolent) or guaranteed (when benevolence cannot be assured) value of the goal function of the principal over all equilibrium states of the agents' game. It should be emphasized that, in general, this stage causes major theoretical difficulties and seems the most time-consuming one for a researcher.

When the set of solutions to the control problem is calculated, one can move to the *fourth stage*, notably studying stability of the solutions. *Stability analysis* implies solving (at the very least) two problems. The first problem is to study the dependence of optimal solutions on parameters of the model; in other words, this is analysis problem for *solution stability* in a classical representation (e.g., well-posed character of an optimization problem, sensitivity, and stability of principles of optimality). The second problem turns out to be specific for mathematical modeling. It consists in theoretical study of *model adequacy* with respect to the real system; such study implies efficiency evaluation for those solutions derived as optimal within the model when they are applied to the real OS (modeling errors cause difference of the model from the real system). The solution of the adequacy problem is a *generalized solution of a control problem*, that is, a family of solutions parameterized by the value of the guaranteed efficiency within a specific set of real OS.

Thus, the aforementioned four stages constitute a general theoretical study of the OS model. To use the results of a theoretical study in practice the model must be tuned (i.e., identified in a series of simulations; these are the *fifth* and the *sixth stage*, respectively). The system is identified using those generalized solutions that rely only upon the information available in reality. In many cases the simulation stage appears necessary due to several reasons. First, one seldom succeeds in obtaining an analytical solution to the optimal control synthesis problem and in studying its dependence on the parameters of the model. In this situation simulation gives a tool to derive and assess the solution. Second, simulation allows verifying validity of hypotheses adopted while constructing and analyzing the model (in the first place, with respect to behavioral principles of system members, e.g., procedures used to eliminate uncertainty and rules of rational choice). In other words, simulation gives additional information on adequacy of the model without conducting a natural experiment. Finally (and this is the third reason), employing business games and simulation models in training lets the managing staff master and test the suggested control mechanisms.

The *seventh stage*, that of implementation, finalizes the process; it includes training of the managing staff, implementation of control mechanisms (designed and analyzed at the previous stages) in real OS, with subsequent efficiency assessment, correction of the model, and so on.

From the description of the general control technology in OS, we pass to the general approaches to solving theoretical control problems in OS.

To proceed to the detailed consideration of control problems, it is necessary to get back to the models of decision making by a controlled subject. Mathematical description of human behavior is provided by decision theory and by game theory. Therefore, let us do a short excursus to these theories to understand what known models could be utilized.

1.2 Models of Decision Making

Construction of a control model for an organizational system requires a behavioral model of people, who form such a system; that is, a model of human decision making is required.

One of the most important steps in creating a formal model of human behavior was the development of a concept of *utility maximization* in economics [99, 116] in the 1850s. This was the concept of an economic person acting to maximize his or her utility. Notwithstanding the limited character of the theory (e.g., it is not always clear what the utility is, and why the person strives to maximize it), the concept turns out to be very fruitful, and till now remains the most convenient and, at the same time, deep enough to model many effects not only in economics but also in politics, management, and psychology.

Suppose there is a subject (an agent) who is able to choose actions from a certain set A. Assume that preferences of the agent are described by a *goal function* $f(y): A \to \Re^1$ (sometimes it is also referred to as a preference function or a payoff function, which is more typical for game-theoretical literature; in the sequel we will use these terms as synonyms), which maps the set A of his or her *actions* (alternatives) into the real axis $\Re^1$. This function measures the level of desirability of an alternative—the greater is the value of the goal function, the more desirable is the action to the agent.* Hence, the agent will maximize his or her gain and will choose the actions from a *choice set*, which represents a certain set of maxima of his or her goal function:

$$P(f(\cdot), A) = \operatorname*{Arg\,max}_{y \in A} f(y).$$

Consequently, the choice set of the agent depends on his or her preferences $f(\cdot)$ and on the set A of feasible actions.

The assumption that the agent chooses an action from the choice set, i.e., seeks to maximize his or her goal function, is referred to as *the hypothesis of rational behavior*; it states that the agent always chooses the best feasible alternative

---

* In this book we use utility functions to describe humans' rational behavior, as it is the simplest and the most convenient modeling approach. At the same time, extensive literature exists on more general models of decision making, for example, in terms of *preference relations* (binary [1, 51, 121, 122] or fuzzy [80, 131]). Moreover, some models take into account bounded rationality [144] or even irrational behavior [79] of decision makers.

(ensuring the maximum to his or her goal function), taking into account the available information [83].

So, if we consider an agent as a controlled subject, we see that within the framework of the considered behavioral model, one can affect the behavior of the agent by, first, imposing influence of his or her goal function, or, second, via an impact on the set of his or her feasible actions. The former type of control is called motivational control, while the latter is an institutional one.

Example 1.1

Let an agent be an industrial enterprise that chooses an action, viz. the production volume $y \in A = [0; y_{max}]$, where y_{max} denotes the maximal possible production under current technology and manufacturing assets. The product of the enterprise is sold at a price $\theta > 0$ per piece, while manufacturing costs constitute $y^2/2\ r$, where the parameter $r > 0$ stands for the "efficiency" of production. The goal function of the enterprise is its gross margin, which equals the difference between total sales and costs:

$$f(y) = \theta\ y - y^2/2\ r.$$

If the enterprise strives to manufacture the amount of products that maximizes the gross margin, it will choose the action

$$y^*(\theta, r, y_{max}) = \min \{y_{max}; \theta\ r\}.$$

The optimal action of the enterprise depends on several factors: the market price θ, the production efficiency r, and the production capacity y_{max}. •*

The aforementioned decision model is simple (probably, even elementary), and in real life one would hardly face situations when the choice of a subject explicitly defines his or her level of satisfaction. Sometimes, however, certain external

* Here and below the symbol • indicates the end of an example or a proof.

factors also influence the outcome along with actions of a decision maker. Let us incorporate them into the model of decision making. Assume a parameter $\theta \in \Omega$ exists, known as a *state of nature*. The preferences of a subject (an agent) depend on, firstly, the choice of his or her action and, secondly, this state of nature. In other words, the preferences are defined on the Cartesian product of the set A of feasible actions and the set Ω of feasible states of nature, while the goal function maps this product into the real axis:

$$f(y,\theta): A \times \Omega \to \Re^1.$$

In contrast to the previous case it seems impossible to suggest a simple formula for the problem of goal function maximization; the reason is that, in general, the action maximizing the agent's goal function depends on a value of the state of nature, which is unknown at the moment of making a decision. Let us introduce a *hypothesis of determinism*, which helps to deal with decision making under uncertain conditions. The hypothesis is that when making a decision, a subject seeks to eliminate uncertainty and, thus, to make a decision under complete information. For this, he or she replaces the goal function being dependent on uncertain factors for the one that depends only on the parameters he or she controls.

Several options are then possible:

1. First of all, the agent can substitute a certain value $\hat{\theta}$ of the state of nature into the goal function and choose the action y to maximize the function $f(y,\hat{\theta})$. However, in many cases choosing the correct value $\hat{\theta}$ seems nontrivial.
2. Suppose the agent is a pessimist believing that the worst-case state of nature takes place. Such decision principle is referred to as the principle of *maximal guaranteed result* (also called the *maximin principle*) and lies in the following. An action of the agent would attain a maximum to his or her goal function provided he or

she expects the worst value of an uncertain parameter. In this case, he or she starts with computing a minimum of the preference function with respect to the state of nature; then he or she evaluates a maximum over his or her action:

$$y^g \in \operatorname{Arg}\max_{y \in A} \min_{\theta \in \Omega} f(y, \theta).$$

An obvious benefit given by the described principle of decision making is that it yields the lower-bound estimate of the goal function value. In other words, we obtain the lower-bound estimate as a reference point. At the same time, the considered principle may seem too pessimistic in real-life situations, and the minimum value is considerably underestimated if nature is considered indifferent to the decision maker (DM).

3. Besides the extreme pessimism maximin principle, there also exists an opposite option of extreme optimism. In particular, one may reckon that nature is favorable to the DM, thus "choosing" the action most preferable to the DM. In this case, the maximum for the goal function should be chosen under implementation of the "best" state of nature:

$$y^o \in \operatorname{Arg}\max_{y \in A} \max_{\theta \in \Omega} f(y, \theta).$$

The discussed principle of decision making is called the *criterion of optimism*; it yields the upper estimate of the agent's payoff. Obviously, the considered principle of uncertainty elimination shares the same shortcoming as the previous one—the extreme optimism and extreme pessimism are rare phenomena in real life!

One can consider various combinations of the criteria introduced. For instance, one may maximize the linear combination of pessimistic and optimistic estimates of the payoff (known as the *Hurwitz criterion*), balancing between the optimism and the pessimism. Alternatively, the pessimism of the minimax criterion could be weakened by minimizing the maximal *regrets*—the loss in payoff compared with the one under the best decision based on the complete information. This "minimum regret" criterion is also known as the Savage criterion (after Leonard J. Savage).

Now, assume we have additional information on the value of the uncertain parameter θ belonging to the set Ω. Let a probability distribution $p(\theta)$ over the mentioned set be known (this sort of uncertainty is called *probabilistic*); then it seems reasonable to utilize the knowledge and eliminate the uncertainty in the following way. We have the goal function, which depends on the actions of the agent, and the value of the uncertain parameter. Taking mathematical expectation with respect to the known distribution, one obtains the function of the *expected utility* $w(y) = \int_{\Omega} f(y,\theta)p(\theta)d\theta$. Eliminating the uncertainty by means of mathematical expectation is one of the most popular ways to come to the situation of deterministic choice. Then the function of expected utility (which no more depends on the environment) is maximized with the choice of the agent's action.

Example 1.2

Within the framework of Example 1.1, suppose there is no capacity constraint ($y_{max} = +\infty$). Let the future value of the market price be uncertain: $\theta \in \Omega = [\theta^-; \theta^+]$. Let $\hat{\theta} \in \Omega$ be the future product price, which is unknown at the moment when the production volume must be chosen.

According to the principle of maximal guaranteed result, the enterprise should expect the price θ^- (since it delivers minimum to the goal function $f(y) = \theta\, y - y^{2/}2r$ under any fixed action of the agent). Choosing the action $y^g = \theta^- r$, the gross margin of the enterprise is

$$f^g = r\,\theta^-(\hat{\theta} - \theta^-/2).$$

Following the optimistic criterion, the enterprise expects the maximum price, chooses the action $y^0 = \theta^+ r$, and makes the gross margin

$$f^o = r\,\theta^+(\hat{\theta} - \theta^+/2).$$

Assume that the enterprise is a priori aware of the market price $\hat{\theta}$ (thus operating in the absence of uncertainty). Then it chooses the action $y^* = \hat{\theta}\, r$ and gains the profit $f^* = r\hat{\theta}^2/2$.

Consider the case of probabilistic uncertainty. For instance, suppose the enterprise knows that the price is distributed uniformly on $[\theta^-; \theta^+]$. Then evaluating the expected value $w(y) = (\theta^- + \theta^+)\, y/2 - y^2/2\, r$, the enterprise would choose the action $y^p = (\theta^- + \theta^+)\, r/2$ and get the profit

$$f^p = \left[\hat{\theta} - (\theta^- + \theta^+)/4\right]\ (\theta^- + \theta^+)\, r/2.$$

Obviously, the presence of uncertainty reduces the payoff of the enterprise (in comparison with the case of complete information). For instance, set $r = 1$, $\theta^+ = 4$, $\theta^- = 1$, $\hat{\theta} = 3$. Then

$$f^g = 2.5;\ f^o = 4;\ f^p = 4.375;\ f^* = 4.5. \bullet$$

There are also other ways to eliminate the uncertainty. One can estimate the *risk*, or the probability of the event that the value of the goal function is smaller than a given quantity. The possible principle of decision making consists in minimizing the risk (under certain conditions this rule is equivalent to maximization of the weighted sum of the mean and the variance of the random payoff). Approaches may differ, but the main idea is the same—to remove the dependence of the goal function on the uncertain parameter, as the hypothesis of determinism requires the uncertainty to be eliminated (with account to all available information!). After that, the decision is made under complete information.

At the end of the section we would like to mention yet another type of uncertainty—a fuzzy one, when the *membership function* is defined on the set of the states of nature, which reflects a level of a priori assurance that a certain value of the environment variable can be the true one. The classic survey of the models of decision making under fuzzy uncertainty can be found in [80].

Now we are ready to pass to the new level of complexity of the decision-making procedures. We have started with a certain function that depends only on the action of the agent; on the next step we introduced the uncertainty of nature in the form of a parameter describing the external environment. However, in social and economic systems, in particular in organizations, an agent often interacts with other agents; therefore, modeling of agent's behavior demands a model of interaction of this sort. Such models are developed and studied using *game theory.*

1.3 Basics of Game Theory

Game theory is the study of decision making in conflict situations, that is, situations of interaction between subjects (agents, players) when the payoff of every subject depends on actions of all participants and these payoffs may differ.

Let us introduce a formal description. Assume there exists a set of *players* $N = \{1, 2, \ldots, n\}$. Player i chooses the action y_i from a set A_i of feasible actions, $i \in N$. Actions of all players are said to be a *strategy profile*: $y = (y_1,\ldots,y_n)$. The *payoff function* of player i depends on the vector y containing actions of all players; it is a certain mapping $f_i(y) : A' \to \Re^1$ of the set representing the Cartesian product of the sets of feasible actions of all players, $A' = \prod_{i \in N} A_i$, into the real axis. In other words, each profile (combination of players' actions) is an outcome resulting in some payoff for every player involved. The tuple $\Gamma_0 = \{N, \{f_i(\cdot)\}_{i \in N}, \{A_i\}_{i \in N}\}$ composed of the set of players (agents), their payoff functions, and the sets of agents' feasible actions is referred to as a *normal-form game* if the players choose their actions once, simultaneously, and independently (i.e., having no opportunity to negotiate or coordinate their strategies or behavior). The introduced model is, in fact, a model of noncooperative behavior.

Let us consider player i and apply the hypothesis of rational behavior to his or her situation of decision making. Being rational, player i chooses the i-th component of the action profile y to maximize his or her payoff: $f_i(y) \to \max_{y_i}$. However, the action ensuring the maximum value of the payoff function depends in general on the actions chosen by the other agents. The situation is similar to the case of nature uncertainty, since the payoff function and, thus, the best action $y_i^*(y_{-i})$ of each agent depends on the actions of the remaining players (his or her *opponents*), that is, on the vector $y_{-i} = (y_1, \ldots, y_{i-1}, y_{i+1}, \ldots, y_n)$. This vector is called an *opponent's action profile* for player (agent) i.

To choose an action, an agent should know the behavior of the remaining agents. Hence, it is necessary to make assumptions regarding the behavior of the other players. We can try to apply the technique used to eliminate uncertainty in the case when the decision is made by a single subject; the opponents' action profile plays the role of the environment parameter for player i. The uncertainty of this sort is called the *game uncertainty* (the one caused by purposeful behavior of the opponents). Since the players are supposed to choose their actions simultaneously and independently, each player chooses his or her action with no information on the actions of the others players. Let us focus on possible alternatives.

1. Suppose player i believes that the remaining players act against him or her. Under this hypothesis of pessimism, player i chooses the *guaranteed* (or *maximin*) *action*

$$y_i^g \in \operatorname{Arg}\max_{y_i \in A_i} \min_{y_{-i} \in A_{-i}} f_i(y_i, y_{-i}),$$

where $A_{-i} = \prod_{j \neq i} A_j$. True actions of other players coincide with this pessimistic forecast only when the other players play (choose their actions) against player *i*, ignoring their own preferences. An evident imperfection of such a decision-making principle is that the player's beliefs on the opponents' actions are inconsistent with the hypothesis of rational behavior of opponents (first of all, each player tries to maximize his or her payoff instead of hurting the opponent). In real life this situation is not widespread, although it can take place in the case of *opposing interests.*

The vector of maximin actions of all players is referred to as a *maximin* (or *guaranteed*) *equilibrium.* This is one of the ways to define the *solution* of the game, its most probable outcome given that the players behave rationally. The game solution concept determines a model of agents' behavior in a conflict situation and, thus, defines the behavioral model of a controlled system, if the set of agents represent a controlled object.

However, this is not the only alternative, and nowadays the major challenge of game theory is the absence of a common (universal) *game solution concept* (i.e., a strategy profile being stable in a certain sense). Different models involve different assumptions leading to different concepts of equilibrium, and we will discuss the most popular of them.

2. Let the payoff function $f_i(y)$ of player *i* be maximized by the same action y_i^d of this player regardless of actions of the remaining players. If this is the case one says that player *i* has a *dominant strategy.* Formally, the strategy y_i^d turns out to be dominant if (irrespective of the opponents' actions) the player's payoff achieves its maximum at the dominant strategy:

$$\forall y_i \in A_i \quad \forall y_{-i} \in A_{-i} \quad f_i(y_i^d, y_{-i}) \geq f_i(y_i, y_{-i}).$$

Note that an arbitrary (yet, the same) opponent's action profile appears in both sides of this inequality.

If each player has a dominant strategy, the profile of these dominant strategies is referred to as a *dominant strategy equilibrium* (DSE) $\{y_i^d\}_{i \in N}$. The situation when the dominant strategy equilibrium exists is ideal for modeling. In this case every player can be considered as making decisions independently, and explaining independent decision making is much easier. Unfortunately, in most economic and management problems DSE does not exist.

3. More typical is the situation when the *Nash equilibrium* (NE) exists. In the early 1950s (see [120]), American mathematician John Forbes Nash suggested the following idea. A vector of agents' actions can be considered as a stable result of agents' interactions when none of the players can benefit by unilaterally changing his action. This means that any agent changing his or her action would not increase his payoff provided the remaining agents keep their actions fixed.

   Let us give a formal definition of the Nash equilibrium. In the game $\Gamma_0 = \{N, \{f_i(\cdot)\}_{i \in N}, \{A_i\}_{i \in N}\}$, a vector $y^N \in A'$ is the Nash equilibrium if and only if for every player i and for any action $y_i \in A_i$ of this player the following inequality holds: $f_i(y_i^N, y_{-i}^N) \geq f_i(y_i, y_{-i}^N)$. In other words, the payoff of player i is maximal when he or she chooses his or her equilibrium strategy provided the remaining players also choose their equilibrium strategies.

Example 1.3

Consider two agents—different departments in an enterprise. Each agent chooses a nonnegative output. Products of every agent are sold on a market for $1 per unit. The cost incurred by an agent depends on the efficiency of his or her production (see the coefficient r in the cost function) and on the output of the other agent. The higher the output of the opponent, the lower the costs of the given agent. The payoff function of agent i, $f_i(y)$, is expressed by the difference between his or her income y_i and the costs

$$c_i(y,\ r_i) = \frac{(y_i)^2}{2(r_i + \alpha y_{3-i})},\, i = 1, 2;$$

here the parameter $\alpha \in [0;\ 1)$ describes the level of agents' interdependence.

Find the optimum from the first order conditions. Evaluate the derivative of the y_i-concave functions

$$f_i(y) = y_i - \frac{(y_i)^2}{2(r_i + \alpha y_{3-i})},\, i = 1, 2,$$

with respect to y_i and equalize the results to zero. Collecting together all n equations for $i = 1, \ldots, n$, and solving the corresponding system subject to actions of agents we obtain the Nash equilibrium of the agents' game:

$$y_1^* = \frac{r_1 + \alpha r_2}{1 - \alpha^2}, \quad y_2^* = \frac{r_2 + \alpha r_1}{1 - \alpha^2}.$$

Obviously, equilibrium actions grow when α grows. •

The stated approaches, DSE and NE, differ in that the notion of dominant strategy equilibrium includes arbitrary opponents' action profile (i.e., a dominant strategy is the best regardless of the profile of opponent actions). At the same time, the Nash strategy appears the best only under the Nash-equilibrium actions of the opponents.

One of the most important advantages of the Nash equilibrium is that the set of equilibria is typically not empty. The drawback is that the Nash equilibrium is not always unique. If you have two equilibria, it is hard to guess which of them will be the outcome of the game of the agents. This case requires introducing additional assumptions on agents' behavior.

Moreover, a Nash equilibrium may be unstable with respect to deviations of two or more players. According to the NE definition, it seems unbeneficial for a single agent to deviate from the equilibrium; nevertheless, it does not mean that, in the case of the agreement between two agents about simultaneous deviation, they would not both get extra payoff. Hence, the Nash equilibrium turns out to be an essentially noncooperative concept of equilibrium.

4. At the end, let us introduce yet another game solution concept—that of Pareto-efficient profiles. Consider a feasible vector $y^P \in A'$ of agents' actions. It is said to be *Pareto efficient* if and only if for any other vector of actions there exists an agent with strictly less payoff than at the Pareto point, i.e, $\forall y \neq y^P \; \exists i \in N$: $f_i(y) < f_i(y^P)$.

   In other words, deviation from a Pareto profile cannot simultaneously increase the payoffs of all players. The idea seems attractive; indeed, it suggests that if we can improve the payoffs of all players, we definitely must do so. It seems that any reasonable game solution concept must suggest only Pareto-efficient outcomes. An interesting question concerns the relation between the aforementioned equilibrium strategies and Pareto efficiency. Notably, one always wishes that an outcome described by individual maxima would be efficient for the society as a whole. Unfortunately, Pareto efficiency is not assured by both of the previously introduced game solution concepts.

Example 1.4

Consider a classical example [56]. Suppose that each of players chooses his or her actions from the set $A_i = [0; 1]$. The payoff of agent i is $f_i(y) = y_i + \sum_{j \neq i}(1 - y_j)$. Let us find the dominant strategy equilibria and the Nash equilibria for this example.

Looking at the shape of the agent's payoff function, one can easily see that it achieves its maximum when the agent chooses the maximal value of his or her action, irrespective of the actions chosen by the remaining players. Indeed, the derivative of the goal function of agent *i* with respect to his or her action is strictly positive regardless of the opponents' action profile. Therefore, every agent would choose maximal value of his or her action (i.e., there exists a dominant strategy for this player). Whatever the opponents do the player under consideration benefits from the increase of his or her action. On the other hand, as the set of feasible actions is bounded, the agent cannot choose the action exceeding the unity; hence, $y_i^d = 1$.

Now evaluate the payoff of every agent in the case of DSE. If all players choose the unity action, they would obtain the same gain: $f_i(y^d) = 1$.

Find the Pareto-efficient vector. In fact, in the case of three or more agents, it is composed of zero actions: $y_i^P = 0$. When all agents choose zero actions, the payoff of agent *i* constitutes $f_i(y^P) = n - 1$. It is impossible to increase simultaneously the payoffs of all agents. Trying to increase the payoff of agent *i*, we start increasing his or her action; however, this, at the same time, reduces the gains of the others since the same action (but with the negative sign) enters the goal functions of the remaining agents.

In the case of three or more players they would earn more from choosing Pareto-efficient actions as compared to the case of dominant strategies so long as $n - 1 > 1$ holds true for $n \geq 3$.

The question is: does a Pareto point represent a Nash equilibrium point (one can easily check that every DSE is a Nash equilibrium)? In other words, is a Pareto point an individually rational strategy? Choosing a nonzero strategy, some player would definitely gain more. Therefore, the player under consideration would increase his or her action up to the unity, and the remaining participants would follow him; consequently, the strategy profile would come down to a DSE being stable, yet beneficial to nobody. •

This example shows that stability against individual deviations is not related in general to Pareto efficiency. The two solution concepts meet in repeated games, where players can negotiate penalties to a player deviating from a collective optimum (i.e., from a Pareto-efficient point). Under a sufficiently high penalty, it appears that each player would (individually and persistently) choose a strategy being beneficial to all agents.

There is another way to deal with the problem of equilibrium inefficiency. For the agents a priori having equal rights, one may appoint a superior (a principal) that would be responsible for their actions (so as the agents avoid deviations and attempts to increase their local gains, but choose a Pareto-efficient action profile). In other words, the principal has the function of preventing agents' deviation from the Pareto optimum. In the case of transferable utility it is even possible to estimate the *price of anarchy*—the amount the agents would like to pay to such a principal for his or her work. This amount equals the difference between the total payoff in the Pareto point and the total payoff under a non-cooperative equilibrium (DSE or a Nash equilibrium). The above arguments provide a game-theoretic basis for the origin of hierarchies.

Hierarchical Games

From the viewpoint of the theory of organization and of control theory, of a major interest are models of games with agents making decisions not simultaneously but sequentially. That is, given a principal and several agents, first the former establishes "rules of the game" and then the latter choose their actions. The games of this sort are referred to as *hierarchical games* or games with a fixed sequence of moves [56, 58].

An elementary model of a hierarchical game is provided by a two-person game: the first player is a principal (P), a leader who moves first; and the second player is a follower, an agent (A) (see Figure 1.11).

Suppose the goal function $\Phi(u,y)$ of the principal and the goal function $f(u,y)$ of the agent both depend on the action $u \in U$ of the principal and on the action $y \in A$ of the agent.

The game Γ_i is where the principal first chooses her action $u \in U$, reports it to the agent, and then the latter chooses his

Figure 1.11 A "principal-agent" basic structure.

action. Analysis of this game reduces to predicting agent's behavior provided the principal's choice is known to the agent.

Find a set of actions attaining a maximum to the goal function of the agent under a fixed choice of the principal: $P(u) = \operatorname{Arg}\max_{y \in A} f(u, y)$. Obviously, this set depends on the action $u \in U$ chosen by the principal. In other words, the principal's action plays the role of a control input, since it impacts a state (in this example, an action) of the agent. Suppose that the principal and the agent are both aware of the goal functions and feasible sets of each other; the former may forecast the reaction of the latter and say, "Being rational, the agent would react to my action by choosing an action from the set of his actions ensuring a maximum to his goal function." What should the principal do to motivate the agent to choose the action necessary to the principal? Being aware of her own payoff $\Phi(u,y)$ (it depends on her action and on an action of the agent), the principal has to identify what action the agent would choose from the set $P(u)$.

The set $P(u)$ may include a single point or several points. In the latter case this uncertainty must be eliminated by introducing an additional assumption on agent's behavior. There exist two standard assumptions of this sort, viz. the criteria of optimism and pessimism (discussed previously). The criterion of optimism claims that as an agent equally valuates all actions from $P(u)$, the principal may adhere to the following reasoning and say, "If the agent does not care what action to choose and is benevolent to me, I can expect him to choose the action that is beneficial to me." Such an assumption meets the principle of optimism in decision theory (see above), and is called the *hypothesis of benevolence.* According to this hypothesis, an agent chooses the action from the set $P(u)$ of the most preferred actions to maximize the principal's goal function (i.e., the action that is the most beneficial to the principal).

If one evaluates a maximum of the function $\Phi(u,y)$ over the agent's actions, the result depends only on the action u of the principal. As a rational player, the principal would choose the action delivering the maximum to her goal function.

Therefore, an optimal "control" (a solution of the hierarchical game) is the action of the principal, which attains a maximum to her payoff $\Phi(u,y)$ over the set of feasible controls provided that the agent chooses the action, most preferable to the principal, from the set $P(u)$ of actions most preferable to the agent given the control u:

$$u^o \in \operatorname{Arg}\max_{u \in U} \max_{y \in P(u)} \Phi(u, y).$$

The pessimistic approach (the maximin principle or a *principle of maximal guaranteed result*) is based on the following arguments of the principal. She says, "As the agent does not care what action to choose from the set $P(u)$, he can choose any, including the worst one, and I have to expect the worst." It leads to the solution

$$u_g \in \operatorname{Arg}\max_{u \in U} \min_{y \in P(u)} \Phi(u, y).$$

In other words, the principal evaluates the minimum of her goal function over the agent's action belonging to the set $P(u)$, with subsequent maximization by a proper choice of her control u.

Thus, we have derived two different solutions of the game. Below the first solution is referred to as a *Stackelberg solution* (in honor of German economist Heinrich von Stackelberg, who pioneered the concept of equilibrium for the leader–follower game in the 1930s). The second solution is called simply a *solution of* the *game* Γ_1.

Consider another game where instead of choosing a certain control action a principal announces a commitment about the control she will apply depending on the action chosen by the agent. This game, having important applications in contract theory, is referred to as Γ_2. In this game the subject of principal's choice (in game-theoretical terms it is called a *strategy*) is a function $u = \hat{u}(y)$ of the agent's action. To solve this game (i.e., to suggest some rational strategy to the principal) one can employ the reasoning similar to the previous case. The principal may expect that, depending on the function chosen, the agent would choose a certain action maximizing his goal function (with the principal's choice substituted), as follows:

$$P(\hat{u}(\cdot)) = \operatorname{Arg}\max_{y \in A} f(\hat{u}(y), y).$$

Knowing this, the principal solves the corresponding problem, for example,

$$\min_{y \in P(\hat{u}(\cdot))} \Phi(\hat{u}(\cdot), y) \to \max_{\hat{u}(\cdot)}.$$

The expression given is a standard formulation of the basic *game-theoretic control problem* in an organizational system. This setting is called *basic*, since the model includes only two decision makers with the goal functions and sets of admissible actions being a *common knowledge*, and no uncertainty.

At the same time, from the mathematical point of view, this optimization problem is far from trivial. The problem is to find a function $u = \hat{u}(y)$ to maximize the functional that depends both on the function and its argument.

The game Γ_2 was studied in detail and solved by Yuriy Germeier in the early 1970s [58]; in particular, he demonstrated that when side payments are allowed (entering additively the goal functions of the players) the optimal strategy of the principal consists of two modes [57]. The first is a *reward mode* (when the agent is paid for choosing the actions desired by the principal). The second is a *punishment mode*—the agent is penalized by the principal when the former chooses the actions that are not beneficial to the latter. This result is widely used in solving incentive problems in organizational systems (see Chapter 3).

Coming to more complex interaction schemes, one may construct the game Γ_3, where the agent chooses the relation of his action on the action of the principal, being aware of the commitment of the principal to choose certain action given a relation the agent chooses. In other words, here the agent's strategy is a function, while the principal's strategy is a function of the latter function. Note we have two scalar-type strategies in the game Γ_1; the game Γ_2 employs strategies in the form of a function and a scalar, a function and a function on function in Γ_3, and so forth.

It is possible to build the game Γ_4, where the strategy of a principal is a composite function ("a function of a function of ..."). From the mathematical point of view the structure of interaction can be complicated infinitely. One may construct games of any (arbitrarily high) order; the only intricacy would be finding a proper interpretation.

The interpretation of the game Γ_3 is simple. The principal promises the agent to provide him with a certain quantity of resource and asks to report his plans of resource utilization depending on the obtained quantity. To drive the agent to the desired action the principal announces the resource allocation rule, which depends on these plans.

The following question arises: does the principal get anything from nested games (from a growing rank of reflexion)? For instance, is the game Γ_{106} more profitable to her than the game Γ_{1015}?

N. Kukushkin proved that all even games of the form Γ_{2k}, where $k = 1, 2, \ldots$, are equivalent to the game Γ_2 (in the sense of the principal's payoff) [86]. At the same time he demonstrated that all odd games Γ_{2k+1}, $k = 1, 2, \ldots$, are equivalent to the game Γ_3. Hence, an infinite family of hierarchical games (having an order greater than three) was actually reduced to two games, Γ_2 and Γ_3. In addition, it was shown that the principal's guaranteed results in these games satisfy the inequality $\Gamma_1 \leq \Gamma_3 \leq \Gamma_2$.

The Kukushkin theorem implies that when a principal can choose a game form to play, she should choose the game Γ_2, as the most beneficial to her and simple. Otherwise, the principal should choose the game Γ_3; in the worst case, she should participate in the game Γ_1. Playing games of the order 4 (or greater) gives no extra benefit.

Games and Structures

Here we briefly introduce a generalization of hierarchical games to the case of more than two players. The described model is based on [27] and allows for description of decision-making processes in multilayer organizations.

Consider the basic idea providing the complete picture and supporting transition from simple problems to their complex counterparts; the latter are reduced to the former and thus become more comprehensible.

From systems analysis literature it is known that an organizational structure is closely related to a hierarchy of decision making that determines the order of making decisions by different agents in an organization. Consider the following scheme (see Figure 1.12). The process of single subject decision making (Figure 1.12a) has been earlier explained on the basis of the hypothesis of rational behavior (HRB). Notably, a rational agent always strives to maximize his or her goal function with a proper choice of the action (the chosen action should attain a maximum of the goal function). Next, we have made the situation more complex by considering several subjects making decisions simultaneously (Figure 1.12b).

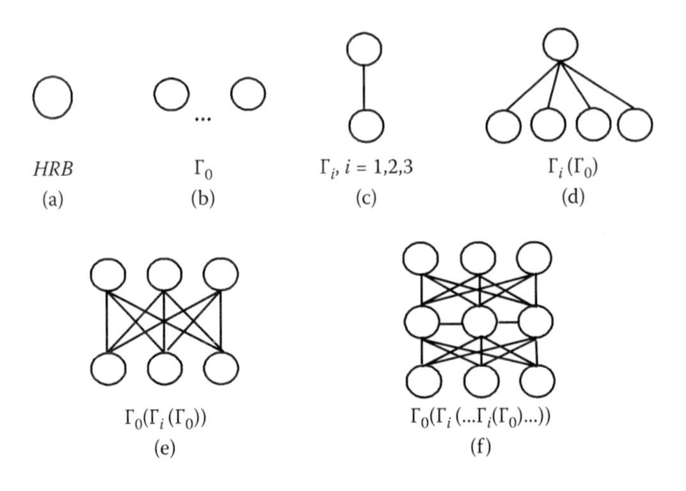

Figure 1.12 Games and structures.

This interaction is interpreted in management as *horizontal coordination* and has been modeled by a normal-form game Γ_0. Finally, we have focused on the situation with two agents interacting vertically (Figure 1.12c). This case stands for a superior–subordinate relation in an organizational structure and is modeled by a hierarchical game Γ_i, $i = 1, 2, 3$.

Consider an organizational structure with single principal and multiple agents (Figure 1.12d). How can it be described in game-theoretic terms? Obviously, interaction of agents at the same level of decision making can be modeled by the game Γ_0. On the other hand, interaction within a principal–agent chain is modeled with the game Γ_i. Hence, such structure can be formally represented by the game Γ_i defined over the game Γ_0. In other words, it makes a hierarchical game, yet it is defined not for a single agent (maximizing his goal function) but for a set of agents, playing their own game.

Consider the more complex structure depicted in Figure 1.12e. There are several principals at the higher level making decisions simultaneously, each announcing a rule of choosing his or her own action according to the game Γ_i, so in general the strategy of the principal is the function of the actions of all agents at the lower level. So, with the agents playing Γ_0 given the strategies of the principals, the principals also engage in the game Γ_0 at their level, actions in this game being the functions of the agents' action profile. The resulting game can be denoted by $\Gamma_0(\Gamma_i(\Gamma_0))$.

In general, an organization structure may constitute several levels, with agents occupying the lowest one, intermediate principals settling the middle levels, with one or several top managers holding the highest level (Figure 1.12f). The principals at the top level choose their actions (in the case of the game Γ_1) or report their commitments to choose a certain action depending on the actions chosen by their immediate or indirect subordinates. Then the principals at the second-top level substitute these actions or commitments in their goal functions and, in turn, choose their strategies as actions or functions of subordinates' actions playing Γ_0 at their level and so forth. The agents at the lowest level of a hierarchy substitute all superiors' strategies into their goal functions, which now depend on the agents' actions only, and play the game Γ_0 to determine their equilibrium actions. Then the system of commitments is rolled up, determining the actions of the principals and, thus, the outcome of the whole game.

The reader interested in a more detailed introduction to game theory can refer to the brilliant textbooks [52, 116].

1.4 Classification of Control Problems in Organizational Systems

In Section 1.1 we discussed several types of control. *Control in an OS*, interpreted as an impact on the controlled OS to ensure its desired behavior, may affect each of six parameters of the OS model. Notably, one can distinguish:

1. Staff control
2. Structure control

3. Institutional control
4. Motivational control (control of preferences and interests)
5. Informational control (control of information being available to OS participants at the moment of decision making)
6. Move sequence control (closely related to the choice of a hierarchical game form discussed in the previous section)

Hence, *the first basis used to classify control problems in OS* is provided by *the subject of control*, notably, a certain component of OS modified in the process of control and as a result of control. Thus, we have already given a classification of control (see prior discussion) based on those components of the controlled system (to be more precise, its model), influenced by control; the list of these components includes the staff, structure, feasible sets, goal functions, and information. Obviously, in general the impact may (and should) be applied to all components mentioned simultaneously. Seeking for an optimal control consists in identification of the most efficient feasible combination of all controllable parameters of an OS.

Nevertheless, the theory of control in organizations [127] traditionally studies a certain system of nested control problems (solutions to "special" problems are widely used to solve more "general" ones). Today, there are two common ways to describe the model of OS (as well as to formulate and solve the corresponding control problems). They are referred to as *bottom-up* and *top-down* approaches.

According to the bottom-up approach, special problems are solved first and then used to state and solve more general ones. An example of a special problem is that of incentive scheme design. Suppose this problem has been solved for any possible staff of OS participants. Next, one may set the problem of staff optimization, that is, that of choosing a certain staff to maximize the efficiency (under a proper optimal incentive scheme). An advantage of this approach is consistency of solutions at different levels. A shortcoming is high complexity due to the large number of possible solutions to the upper-level problem, each requiring the solution of the corresponding set of special problems.

The top-down approach does not have this shortcoming; this approach states that general upper-level problems must be solved first, while their solutions serve as constraints for the special lower-level problems. Such decomposition technique is widely used in practice of technology and organization design. Typically, when creating a new department, a top manager of a large-scale company will not elaborate the details of employees' interactions in this department. Instead he or she will delegate this duty to the head of the department being created. The task of the top manager is to provide the head of the department with necessary resources and authorities. The trade-off is that this approach lacks global optimality, as the consequences of higher-level decisions are not traced to the end.

Construction of an efficient control system for an organization requires combining both approaches in theory and in applications.

Let us continue classification of control in organizational systems. Possible extensions of the *basic model* of the control problem discussed previously (see Figure 1.11 and Figure 1.12c) can be constructed by introducing:

1. *Dynamics* (a principal and an agent make decisions repeatedly); this is an extension with respect to the subject of move sequence control.
2. *Multiple agents* (i.e., several agents make decisions simultaneously and independently); this is an extension with respect to the subject of staff control.
3. *Multiple layers* (the system includes three or even more levels of hierarchy); this is an extension with respect to the subject of structure control.

4. *Multiple principals* (in particular, common agency problems are considered where several principals share the same agent); this is an extension with respect to the subject of structure control.
5. *Mutual constraints* (when the set of admissible actions of one agent depends on the choice of another agent or a principal); this is an extension with respect to the subject of institutional control.
6. *Incomplete information* (a principal and an agent have incomplete but the same information on essential parameters of the system); this is an extension with respect to the subject of information control.
7. *Private information* (a principal and an agent enjoy different information about environmental parameters); this extension to the subject of informational control is the central matter of mechanism design. In the case of *asymmetric information* control assumes revelation of private information by agents and/or by a principal.

Therefore, the *second basis* of the classification system is related to the extension of the basic model—that is, the presence or absence of the following features:

1. Dynamics
2. Multiple agents
3. Multiple levels
4. Multiple principals
5. Uncertainty of the future
6. Joint action constraints
7. Revelation of private information

Looking for the elements of these extensions in a real-world problem helps choosing an adequate modeling approach, as every extension mentioned gives rise to different modeling issues and is supported by the extensive literature.

Modeling approaches provide the third basis of classification. Most models of organizational control are either *optimization models* or *game-theoretic models.*[*] Optimization-based models could be further divided into models involving the following tools: *probability theory* (e.g., reliability theory, queuing theory, and statistical-decision theory), *operations research* (e.g., linear and nonlinear programming, stochastic programming, and integer or dynamic programming), *theory of differential equations, optimal control theory,* and *discrete mathematics,* generally, graph theory (results in, e.g., transportation and assignment, scheduling problems, location problem, and resource distribution problem for networks).

Similarly, the models based on a game-theoretic approach are subdivided into the ones involving different game-theoretic frameworks: *noncooperative games, cooperative games, repeated games, hierarchical games, network formation games,* or *reflexive games.*

The *fourth basis* of the classification system of control problems in OS is provided by *control functions* that have to be implemented. In Section 1.1 we mentioned four basic functions of control—planning, organizing, motivating, and controlling.

The *fifth basis* of the control models classification gives the link of the theory to the typical management policies by introducing typical *control issues* (or *typical business cases*) solved with a certain *control mechanism* (a management decision procedure). The mechanisms listed in Table 1.2 are historically identified by the theory and represent so-called key words. Moreover, these problems have well-developed theoretical and experimental grounds, and were successfully implemented in industry.

Let us emphasize that the classification illustrated by Table 1.2 turns out rather formal due to several reasons. On one hand, the values of the classification attributes are, in fact, the

---

[*] The subject of optimization models is searching for optimal values of the controlled parameters of the system (i.e., feasible values being the best in the sense of a given criterion). In game-theoretic models, some of these values are *chosen* by members of the system, possessing personal interests; therefore, the control problem lies in defining rules of the game, ensuring that the controlled agents choose the desired actions.

Table 1.2 Control Functions and Control Mechanisms

Control Functions	*Control Mechanisms*
Planning	Resource allocation mechanisms Mechanisms of active expert assessment Transfer pricing mechanisms Rank-order tournaments Exchange mechanisms
Organizing	Mechanisms of joint financing Cost-saving mechanisms Mechanism of cost-benefit analysis Mechanisms of self-financing Mechanisms of insurance Mechanisms for production cycle optimization Mechanisms of assignment
Motivating	Individual incentive schemes Collective incentive schemes Unified incentive schemes Team incentive mechanisms Incentive schemes for matrix organizational structures
Controlling	Integrated rating mechanisms Mechanisms of consent Multichannel mechanisms Mechanisms of contract renegotiation

classes of control mechanisms that have been analyzed in detail [19]. On the other hand, the same class of the mechanisms may be used to implement several different control functions.

The *sixth basis* of the classification system for control problems in an OS is, in fact, the size of real systems where a certain problem generally arises (e.g., country–region–enterprise–structural department within an enterprise–collective–individual).

The *seventh basis* is represented by a sphere of activity (e.g., a state government, a municipal government, staple industries, construction industry, or service industries). The latter two classifications are important mostly from the practical point of view, as implementation of a basic resource allocation mechanism on the state level or at a single enterprise substantially differs due to legal restrictions and different project organization issues.

At the end let us emphasize the following aspects. On one hand, the suggested bases and values of the classification attributes—notably (1) a subject of control, (2) an extension of the basic model, (3) a modeling method, (4) a control function, (5) a control issue, (6) a scale of real systems, and (7) a sphere of activity—make it possible to give a uniform description both for concrete control mechanisms and their combinations (complexes of control mechanisms). On the other hand, sometimes it might be difficult to classify a specific mechanism; in many cases the same mechanism solves different control problems and is used in various applications.

In the next chapters of the textbook we address (in greater detail) the mechanisms of motivational control (Chapters 3 and 4), informational control (Chapter 5), and structure control in organizational systems (Chapter 6). Mechanisms of *institutional control* and *staff control* are explained in greater detail in [127].

TASKS AND EXERCISES*

1.1. ("The farmers problem, or a Cournot oligopoly"). There are n players ($N = \{1,...,n\}$) with the goal functions $f_i(x) = x_i(n\,X - \sum_{j \in N} x_j)$, $x_i \in [0, +\infty)$, $i \in N$.

1.1.1. Find all pure strategy Nash equilibria provided that $n = 2$.

---

* Some exercises and tasks include references; these are the works answering the corresponding question (alternatively, providing a solution to the corresponding task) or additional information. The tasks possessing an advanced level of sophistication are marked by an asterisk.

1.1.2. Find all pure strategy Nash equilibria for an arbitrary number of players.

1.1.3. Find all Pareto-optimal action profiles provided that $n = 2$.

Hint: The set of Pareto-optimal action profiles coincides with the set of points ensuring maximum to the function

$$\alpha\ f_1(x_1, x_2) + (1 - \alpha) f_2(x_1, x_2),$$

where $\alpha \in [0; 1]$.

1.1.4. Find all Pareto-optimal strategy profiles under an arbitrary number of players.

1.1.5. Compare the total payoffs of the players in the Nash equilibria and at the Pareto-optimal points.

1.1.6*. Evaluate the following limits as n tends to infinity: equilibrium strategies of the players, their payoffs, total equilibrium gains, and total Pareto-optimal payoffs.

This example could illustrate application of the hypothesis of weak impact (see below).

1.1.7*. Demonstrate that for an arbitrary game (with concave and smooth goal functions) the set of Pareto-optimal strategy profiles coincides with the set of points ensuring the maximum of the function $\sum_{i \in N} \alpha_i f_i(x)$ under different values of $\alpha_i \in [0; 1]$ such that $\sum_{i \in N} \alpha_i = 1$.

Hint: Use the definition of Pareto optimality.

1.1.8*. Does the game under consideration allow for Nash mixed strategy equilibria if $n = 2$?

Hint: Use the fact that the goal function of a player is concave in his action and is linear in the strategies of opponents.

1.2. Consider an n-person game with the goal functions

$$f_i(x) = \alpha x_i - \beta \sum_{j \in N} x_j$$

and the strategies $x_i \in [0,1]$.

1.2.1. Find all pure strategy Nash equilibria.
1.2.2. Find all dominant strategy equilibria.
1.2.3. Find all Pareto-optimal strategy profiles.
1.2.4. What are the values of α and β that guarantee Pareto-optimality of a Nash equilibrium?

1.3. Consider a normal-form two-person game with the goal functions

$$f_1(x_1, x_2) = \begin{cases} 2 - x_1, & \text{if} \quad x_1 + x_2 \geq 0.8, \\ 0, & \text{otherwise,} \end{cases} \qquad x_1 \in [0;\, 1];$$

$$f_1(x_1, x_2) = \begin{cases} 2 - x_2, & \text{if} \quad x_1 + x_2 \geq 0.8, \\ 0, & \text{otherwise,} \end{cases} \qquad x_2 \in [0;\, 1].$$

Eliminate dominated strategies. Construct a set of Pareto-undominated strategy profiles. Find a dominant strategy equilibrium or prove that it does not exist.

1.4. ("The Edgeworth box") [99]. Two players are endowed with resources of two types; each player may exchange only a specific type of the resources. The initial quantities of the resources are $x_1^0 = 1,\ x_2^0 = 0$ for player 1 and $y_1^0 = 0,\ y_2^0 = 1$ for player 2. The goal function of a player depends on the volumes of resources owned and is defined by $f_1(x_1, x_2) = x_1(x_2 + 0.1)$ for player 1 and $f_2(y_1, y_2) = (y_1 + 0.1)\, y_2$ for player 2.

1.4.1. Find a bargaining set and a contract curve (i.e., the set of Pareto-optimal exchange strategy profiles; see Section 4.6).
1.4.2. Find a Walrasian equilibrium of the game (a point where the price line simultaneously touches the indifference curves of both players).
1.4.3. Find the set of Nash equilibria for the following game: both players simultaneously announce a quantity of resources they suggest to exchange, and the deal takes place if the offers coincide.
1.4.4. Find Stackelberg equilibria for the following game: at the beginning, player 1 offers the quantity of resources to-be-exchanged, and then player 2 accepts or rejects the offer (if the offer is rejected, no exchange takes place).
1.4.5. Find Stackelberg equilibria for the following game: at the beginning, player 1 announces the price, and then player 2 suggests the quantity of the first resource for exchange at the announced price.
1.4.6*. For the given formulation of the problem, design a game with a unique Nash equilibrium that coincides with a Walrasian equilibrium (see Task 1.4.2).

1.5. For a normal-form game with the matrix

$$\begin{matrix} ({}_1\backslash{}^2) & \begin{pmatrix} x_2 & y_2 \end{pmatrix} \\ \begin{pmatrix} x_1 \\ y_1 \end{pmatrix} & \begin{pmatrix} (7;1) & (0;0) \\ (4;4) & (1;5) \end{pmatrix} \end{matrix}$$

find all Nash mixed strategy equilibria.

1.6. For a normal-form game with the payoff functions

$$f_1 = (x_1 + x_2)^2, \quad x_1 \in [-1;\ 1],$$

$$f_2 = -(x_2 - x_1)^2, x_2 \in [-1;\ 1],$$

find all Nash pure strategy equilibria.

1.7*. For Task 1.6, give an example of Nash mixed strategy equilibrium, which differs from those already found.

1.8*. Give the definitions and illustrative examples for the following terms [116, 125, 127]:

Organization

Organizational system

Mechanism of operating

Control mechanism

Modeling

Activity

Motive

Purpose

Technology

Control

Input–output model

Control efficiency

Control classes

Control types

Control functions

Control methods

Forms of control

Technology of control

Hypothesis of rational behavior

Hypothesis of deterministic behavior

Hypothesis of benevolence

Game

Opponent's action profile

Principle of maximal guaranteed result

Dominant strategy equilibrium

Nash equilibrium

Pareto efficiency
Hierarchical game
Stackelberg equilibrium
Dynamic organizational system
Multi-agent organizational system
Multilevel organizational system

Chapter 2

Examples of Control Mechanisms

In this chapter we consider several simple control mechanisms in organizations, which serve as a management tool to struggle with incentive incompatibility planning and revenue allocation mechanisms. The chapter is a bit of a retrospective, as most of the mechanisms considered were suggested in the early 1980s, but even nowadays these classic mechanisms are used in corporate and public governance. We theoretically estimate their efficiency and provide examples of efficient mechanisms.

2.1 Planning Mechanisms

Distribution of Corporate Orders [20]

Consider a corporation consisting of n enterprises (Ent), referred to as agents. An elementary corporate structure is illustrated in Figure 2.1. Corporate management is performed by a head office, referred to as principal. Functions of principal include establishing corporate mechanisms,

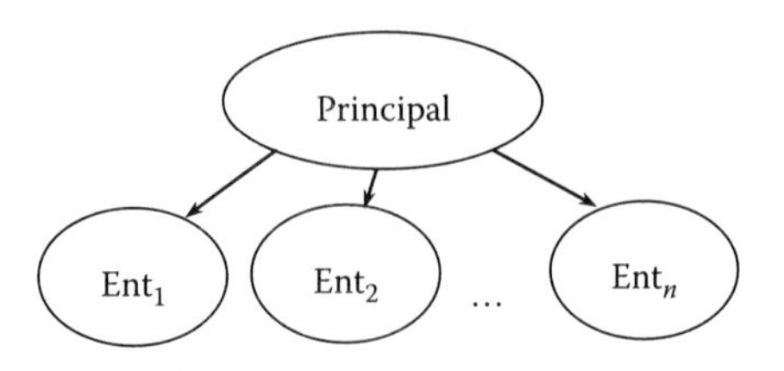

Figure 2.1 A corporate structure.

designing a corporate development strategy, distributing corporate orders, and allocating corporate finance.

Let us focus on a *planning problem*, that is, distribution of corporate orders. Consider a corporate plan for a certain good produced by this corporation; the plan is R at a contracted price of C. The good can be manufactured by any enterprise of the corporation (so, we consider a *horizontally integrated corporation*). The internal planning problem is to distribute the corporate plan among the enterprises to maximize the total profit of the corporation. Introduce the following notation: x_i is a manufacturing plan for the enterprise i, and $\varphi_i(x_i)$ is a manufacturing cost function that depends on the manufacturing volume x_i. In many cases the type of a manufacturing cost function is not of great importance for the analysis of control mechanisms. For simplicity consider the quadratic cost function (see also the model in Section 4.4):

$$\varphi_i(x_i) = \frac{x_i^2}{2r_i}, \quad i = \overline{1,n}. \tag{2.1}$$

Here the parameter r_i stands for the rate of manufacturing efficiency of enterprise i. This function is the simplest example of the manufacturing cost function, increasing and convex in the output, and decreasing in manufacturing efficiency r_i. The profit (value of the payoff function) of enterprise i is written as

$$f_i = C\,x_i - \frac{x_i^2}{2r_i}, \quad i = \overline{1,n}, \tag{2.2}$$

while the total profit of the corporation is

$$\Phi = \sum_{i=1}^{n} f_i = CR - \sum_{i=1}^{n} \frac{x_i^2}{2r_i}, \tag{2.3}$$

as far as

$$\sum_{i=1}^{n} x_i = R. \tag{2.4}$$

The contract price C and the corporate plan R are supposed to be fixed; hence, the maximization problem for corporate profit is reduced to the minimization problem for total costs

$$Z = \sum_{i=1}^{n} \frac{x_i^2}{2r_i} \tag{2.5}$$

under the constraint (2.4).

The optimal solution is provided by

$$x_i^0 = \frac{r_i}{H} R, \quad i = \overline{1,n}, \tag{2.6}$$

with $H = \sum_{i=1}^{n} r_i$. Notably, the total plan must be distributed among enterprises proportionally to their rates of manufacturing efficiency.

However, a problem arises when the principal does not know exact values of efficiency parameters $\{r_i\}$. For instance, the only information available is their possible range $[d; D]$; this uncertainty must be eliminated somehow in the course of decision making. A natural approach is to request the information on the efficiency rates from the enterprises (under the assumption that enterprises possess accurate estimates of the efficiency rates). The described way of receiving the necessary information is said to be a *counter-information flow*. Denote by s_i an estimate of the rate r_i reported by enterprise i to the principal. This estimate is then substituted into the planning rule (2.6), that is,

$$x_i = \frac{s_i}{S} R, i = \overline{1,n}, \tag{2.7}$$

where $S = \sum_{i=1}^{n} s_i$. The following question is immediate: what estimate s_i is reported if every enterprise strives to maximize its profit? The profit is defined by

$$f_i = Cx_i - \frac{x_i^2}{2r_i} = C\frac{s_i}{S}R - \frac{1}{2r_i}\left(\frac{s_i}{S}\right)^2 R^2, \quad i = \overline{1,n}. \tag{2.8}$$

Find the report v_i, which maximizes the profit of the enterprise. From the first-order necessary condition of optimum, we find the report v_i, which makes zero the first derivative of the expression (2.2):

$$v_i = Cr_i, \quad i = \overline{1,n}. \tag{2.9}$$

Assume that $\sum_{i=1}^{n} v_i = CH > R$. This means that the sum of optimal local plans exceeds the corporate plan R. If each enterprise reports the true estimate $s_i = r_i$, we have

$$x_i = r_i\ R/H < Cr_i = v_i, \quad i = \overline{1,n}.$$

In other words, every enterprise obtains a plan smaller than the optimal one.

Naturally, in this case a tendency of overstating the reported estimates appears. If $CH >> R$, then under conditions of Nash equilibrium each enterprise reports the maximal estimate $s_i = D$; this leads to $x_i = R/n$, i.e., the order is equally distributed among the enterprises. The resulting profit of the corporation makes

$$\Phi = CR - \sum_{i=1}^{n} \frac{R^2}{2r_i n^2}.$$

Evidently, it might be appreciably smaller than the profit $\Phi_{max} = CR - R^2/(2H)$ achieved with the optimal plan.

Example 2.1

Set $n = 2$, $r_1 = 3$, $r_2 = 7$, $d = 3$, $D = 7$, $R = 100$, $C = 20$. Let us evaluate the optimal plan and the corresponding profit:

$$x_1^0 = 30,\ x_2^0 = 70, \quad \Phi_{max} = 1500.$$

In the case of Nash equilibrium, we obtain

$$s_1^* = s_2^* = 7, \quad x_i = x_2 = 50.$$

The corporate profit is $\Phi \approx 1400$; hence, the loss is approximately 7%. •

How could planning efficiency be improved? Introduce an *internal price* (a *corporate price,* or a *transfer price*) *of the product.* In fact, this is the price used by the principal "to buy" the good from the enterprises; denote this price with λ. The profit of an enterprise under the transfer price λ is, then,

$$\pi_i = \lambda\, x_i - \frac{x_i^2}{2r_i}, \quad i = \overline{1,n}, \tag{2.10}$$

and attains its maximum under the plan

$$x_i = \lambda\, r_i, \quad i = \overline{1,n}. \tag{2.11}$$

Choose λ such that the sum of profitable plans (under the price λ) equals the corporate plan; notably, use the condition $\sum_{i=1}^{n} x_i = \lambda H = R$ to evaluate $\lambda = R/H$.

Since H is unknown to the principal, we utilize the sum of the estimates S instead of H:

$$\lambda = R/S. \tag{2.12}$$

It should be underlined that the notion of the transfer price bears no relation to real money; a transfer price is a management indicator used to calculate the profit distribution to the manufacturing enterprises. The corporate profit is distributed in direct proportion to the transfer profits:

$$f_i = \frac{\pi_i}{\sum_{j=1}^{n} \pi_j} \Phi_0, \tag{2.13}$$

where Φ_0 stands for the actual profit of the corporation. Equations (2.11)–(2.13) determine a new planning mechanism that differs from the previous one in the following aspects: it involves transfer prices and distributes actual profit in direct proportion to transfer profits.

To assess efficiency of this mechanism, substitute (2.11) and (2.12) in (2.10) and then in (2.13) to get

$$\frac{\lambda^2 \left(s_i - \dfrac{s_i^2}{2r_i} \right)}{\sum_{j=1}^{n} \lambda^2 \left(s_j - \dfrac{s_j^2}{2r_j} \right)} = \frac{\delta_i}{\sum_{j=1}^{n} \delta_j} \Phi_0, \tag{2.14}$$

where $\delta_i = s_i(1 - s_i^2/(2r_i)),\ i = \overline{1,n}$.

Note that (2.14) increases in δ_i. Hence, the maxima of f_i and δ_i are attained simultaneously. On the other hand, the maximum of δ_i is ensured by $s_i = r_i$ (every enterprise reports the true estimate of the efficiency rate). The considered mechanism thereby appears *incentive compatible* (see Chapter 4); that is, under this mechanism the enterprises benefit from truth-telling. The only shortcoming consists in profit redistribution, which could excite dissatisfaction of the enterprises (when a part of their profit is given to the other enterprises). However, there is no profit redistribution for the manufacturing costs functions under consideration. Indeed, the profit of enterprise i is $Cx_i - x_i^2/2\,r_i = (r_i/H)(CR - R^2/(2H))$. After the redistribution, it would be equal to

$$\frac{\delta_i}{\sum_{j=1}^{n} \delta_j}\left(CR - \frac{R^2}{2H}\right) = \frac{r_i}{H}\left(CR - \frac{R^2}{2H}\right).$$

In other words, the profit does not change.

Let us summarize the results. The planning mechanism proposed has three remarkable features:

1. Each enterprise reports the true information about the parameters of its manufacturing costs. Truth-telling appears a dominant strategy for every enterprise.
2. Principal sets the optimal manufacturing plans to implement the corporate plan.
3. No profit redistribution between enterprises takes place.

2.2 Taxation and Pricing Mechanisms

Taxes are usually thought of as belonging to state or local government, as a tool of profits redistribution, and as a source of common projects financing. But they appear to be also a simple and flexible tool of corporate management, helping to decentralize decisions to the lower-level units of an organizational hierarchy (where relevant information is available) while keeping them coordinated and focused toward corporate goals.

Taxation mechanisms [27] serve to determine a share in a profit (an income, revenue), returned by a production unit (an enterprise) to the corporate center (the principal) in the form of an internal tax.* This share is referred to as a *tax rate.* An elementary example is a *flat sharing rule.* Let us introduce the following notation: B is the revenue, Z is the cost, $\Phi = B - Z$ is the income before taxes, IBT, H is the tax amount, and α is the tax rate. Then the tax is calculated as

$$H = \alpha\,\Phi = \alpha\,(B - Z). \tag{2.15}$$

The flat sharing rule motivates production units to manufacture with lower costs and to sell at a higher price (the *cheap–expensive* principle). At the same time, in the case of a monopoly it leads to overpriced products.

In the case of a state-controlled corporation, the principal may be interested in lower consumer prices. To cut prices he or she can impose constraints on product profitability. For example, the principal may decide to withdraw all excess profit when the net profit ratio (calculated as $(B - Z)/Z$) is increasing above the given threshold P_0, and even to impose a penalty to the manufacturer for exceeding the profitability limit. In fact, this malpractice leads to the enterprise working at its threshold level of profitability and selling at the maximal price. Trying to achieve the required net profit ratio, the enterprise artificially overrates its costs. The necessary level of costs Z is evaluated from

$$P_0 = (V - Z)/Z, \tag{2.16}$$

* In the case of a joint venture this tax can take the form of dividends to shareholders, but the same idea can be used in a much broader variety of practical situations, not limited by the strict bound of corporate legislations.

where V is the sales income at the maximal price. Thus, one obtains

$$Z = V/(1 + P_0). \tag{2.17}$$

Therefore, the previous taxation scheme motivates an enterprise to sell at a higher price and manufacture with higher costs (an *expensive–expensive* principle). The question is whether a taxation mechanism exists that brings together the positive features of the presented schemes while not inheriting their shortcomings. The first scheme motivates costs reduction, which is good. However, it is accompanied with inappropriately high prices and growing excess profits of a monopolist leading to inflation and dramatic income inequality (which is bad). The second scheme allows for balancing the prices and costs reducing the degree of wealth inequality, which is its advantage. At the same time, the production becomes inefficient (too costly), which is an obvious drawback. The society undoubtedly would benefit most from a taxation mechanism implementing the *cheap–cheap* principle, that is, the mechanism that motivates an enterprise to manufacture products with lower costs and to sell them at lower prices. Such mechanisms are known as *counter-expensive mechanisms.* The underlying idea is to replace a fixed profitability threshold with that dependent on a *productivity rate.* The productivity rate is defined by dividing "product effect" l by the cost C:

$$E = l/C. \tag{2.18}$$

Define the notion of the *product effect,* which measures the customer value of the manufactured product. Here and in the sequel we will assume the simplest case of inelastic demand, understanding the effect as the revenue evaluated at the limit price l (i.e., a maximal price at which the product is bought by consumers).

Apparently, the net profit ratio norm ρ must increase when the productivity rate increases. Moreover, for an income-sharing rule to be counter-expensive, the income $\pi = \rho(E)\ C$ must decrease with the costs (the smaller the costs, the greater the income). On the other part, the price $P = (1 + \rho)\ C$ should increase with the costs (the lower are the costs, the smaller is the price). One can rewrite the first condition as

$$\frac{d\pi}{dC} = \frac{d}{dC}\left[\rho\left(\frac{I}{C}\right)C\right] = \rho(E) - E\frac{d\rho(E)}{dE} < 0.$$

Similarly, the second condition takes the form

$$\frac{dP}{dC} = \frac{d}{dC}\left[1 + \rho\left(\frac{I}{C}\right)C\right] = 1 + \rho(E) - E\frac{d\rho(E)}{dE} > 0.$$

Both inequalities are easily combined as follows:

$$0 < E\frac{d\rho(E)}{dE} - \rho(E) < 1. \tag{2.19}$$

Let us introduce the shorthand notation $h(E) = E\ d\rho(E)/dE - \rho(E)$; then inequalities (2.19) are rewritten in the form of the following differential equation:

$$E\frac{d\rho(E)}{dE} - \rho(E) = h(E), \tag{2.20}$$

where $h(E)$ is an arbitrary function taking its values within the interval (0;1). This differential equation is easily solved. Consider the function $u(E) = \rho(E)/E$; consequently, $\rho(E) = E\ u(E)$ and $d\rho(E)/dE = u(E) + E\ du(E)/dE$. Substituting it into equation (2.20) one obtains

$$\frac{du(E)}{dE} = h(E)/E^2, \quad u(E) = \int_1^E \frac{h(y)}{y^2}\,dy.$$

In the previous formulas, the additional condition $\rho(1) = 0$ has been employed. It means that when the effect of a product is equal to the manufacturing costs, the income equals zero. Thus, we have derived the function $\rho(E)$, which ensures the counter-expensive property (subject to the income) of a pricing mechanism:

$$\rho(E) = E\int_1^E \frac{h(y)}{y^2}\,dy.$$

Example 2.2

Let $h(E) = k$, $0 < k < 1$. In this elementary case, we obtain the following dependence of the normative net profit ratio ρ on the product effect E:

$$\rho(E) = E\int_1^E \frac{k}{y^2}\,dy = k(E-1).$$

The price is determined by the formulae $P = [1 + k\,(E-1)]\,C$, while the income makes $\pi = P - C = k\,(l - C)$.

Apparently, decreasing C results in a price drop, while the income goes up. Note that the difference $l - C$ defines a net income. A certain share of it (denoted by k) is left to the enterprise; the rest supports growth of customer profits. The function $h(E)$ is chosen according to the following considerations. Recall that $d\pi/dC = h(E)$ and $dP/dC = 1 - h(E)$; hence, the closer $h(E)$ to the zero point, the stronger the impact of the reduced costs on the price drop (the weaker is the influence of the reduced costs on the profit growth). And vice versa, as $h(E)$ approaches the unity, the impact of reduced costs on the price drop relaxes (however, the influence of the reduced costs on the profit growth intensifies). To balance both tendencies one is advised to take $h(E) \approx 1/2$. •

Another strategy is also possible when, under high costs (low productivity), an enterprise is naturally motivated to reduce the costs. Implementation of this strategy requires choosing $h(E)$ close to the unity. On the contrary, under high productivity it seems reasonable to motivate price drops, thus choosing $h(E)$ close to zero. For instance, these requirements are satisfied with the function $h(E) = 1/E$. In this case one obtains

$$\rho(E) = E\int_{1}^{E}\frac{dy}{y^3} = (E - 1/E)/2,$$

$$P = [1 + (E - 1/E)/2]\, C = C + (l - C^2/l)/2,$$

$$\pi = (l - C^2/l)/2.$$

The described principle of counter-expensive mechanisms design can also be applied to *pricing mechanisms*. In fact, setting the price schedule in the form $P = C\,(1 + \rho(E))$, $E = l/C$, where l stands for the limit price, C indicates the net cost, and $\rho(E)$ meets the counter-expensive condition (2.19). Then reducing the net cost increases the income and decreases the price.

2.3 Multichannel Mechanisms

Process automated control systems (PACS) widely adopt the *operator-adviser systems*. These are computer programs that simulate industrial processes and (after a learning period) generate process control recommendations to the operator. In practice, efficiency of such passive advisors often appears to be low. This is because during the learning period the recommendations of the computer program are often inappropriate; as a result, experienced operators ignore them, although with the lapse of time the control actions suggested by the advisor system in situations of normal operation become better than the actions chosen by operators. So the solution of this problem is to design a certain mechanism that would motivate the operators to take into account the advisor's recommendations. A special model (known as a *verification model*) of a technology (e.g., a plant, an appliance, or a device) must be developed for the considered technological process, which forms the core of this motivation mechanism. The model uses the realized process output to restore retrospectively the process output that would be achieved if the operator followed recommendations of the adviser system. When the recommendations lead to the better output (compared with the one obtained under the operator's control), the operator is penalized and is rewarded otherwise. Actually, it inspires a competition between the operator and the advisor system. The experience of implementation of such mechanisms shows that the things change for the better as soon as such active advisors are employed. In many cases, the operators followed recommendations of the model (especially in normal operation modes). Implementation of the two-channel mechanisms in ferrous metallurgy yielded a substantial economic effect [8].

The described two-channel mechanism could be generalized in several directions. First, one may employ several advising channels (*multichannel mechanisms*), for example based on various models and simulation methods. Second, such active advisors could be applied not only to PACS but also to several semi-automated business processes in organizations.

2.4 Incentive Mechanisms for Cost Reduction

Consider an enterprise that consists of n departments. A typical management problem is that of cost reduction. Denote with R the desired total amount of cost reduction. Denote by x_i the plan of cost reduction for department i. Cost-reducing activities require some investments. Let

$$Z_i = \varphi_i(x_i), \tag{2.21}$$

be the investments of department i, incurred by its activity to reduce the costs at the value of x_i.

For example, set

$$\varphi_i(x_i) = x_i^2/(2r_i), \quad i = \overline{1,n}. \tag{2.22}$$

Consider the following incentive mechanism for cost reduction (note it resembles the mechanism of order allocation in a corporation; see Section 2.1). The department receives financial resources h_i from a centralized fund in direct proportion to x_i, that is,

$$h_i = \lambda\, x_i, \quad i = \overline{1,n}. \tag{2.23}$$

Here λ stands for a price of cost-reducing efforts, common for all departments.

A cost reduction plan is designed on the basis of cost function estimates reported by the departments. Suppose each department reports the estimate s_i of the rate r_i in the cost function (2.22). The plan $x = \{x_i\}$ is defined by

$$x_i = \frac{s_i}{S} R, \quad i = \overline{1,n}. \tag{2.24}$$

In this formula, $S = \sum_{j=1}^{n} s_j$, and the value of the price λ is chosen to fulfill the total cost reduction plan R, that is,

$$\lambda = R/S. \tag{2.25}$$

Let us analyze the mechanism described. Assume that goal functions of the departments are given by the difference between the financing received from the centralized fund and their investments in the cost reduction activities to achieve at the planned amount x_i of cost reduction:

$$f_i = h_i - Z_i = \lambda\, x_i - x_i^2/(2r_i), \quad i = \overline{1,n}. \tag{2.26}$$

Substituting expressions (2.24) and (2.25) into (2.26), we have

$$f_i = \lambda^2 \; s_i \left(1 - \frac{s_i}{2r_i}\right) = \left(R/S\right)^2 \; s_i \left(1 - \frac{s_i}{2r_i}\right), \quad i = \overline{1,n}. \tag{2.27}$$

Under a sufficiently large number of departments, the impact of the estimate s_i (provided by the i-th department) on the price λ is relatively small. Hence, the assumption that the enterprise does not account for such impact when reporting the estimate s_i (known as the *hypothesis of a weak impact* [23, 30, 127]) seems rather reasonable. In this case, the maximum of the goal function (2.27) is ensured by the department through reporting the estimate $s_i = r_i, i = \overline{1,n}$; in other words, the enterprises truthfully report their private information about their cost reduction abilities to the principal.

The hypothesis of weak impact does not hold if there exists an enterprise with a relatively big rate r_i. Suppose this is enterprise 1; in addition, assume that $r_1 > H - r_1$, i.e., $r_1 > H/2$. The hypothesis of weak impact turns out relevant for the remaining enterprises in a corporation; therefore, $s_i = r_i$, $i \neq 1$. Denote $H_1 = H - r_1$.

The goal function of Enterprise 1 becomes equal to

$$f_1 = \left(\frac{R}{s_1 + H_1}\right)^2 \left(1 - \frac{s_1}{2r_1}\right) s_1.$$

Maximizing over s_1, we obtain

$$s_1 = \frac{H_1 \; r_1}{H_1 + r_1}. \tag{2.28}$$

If $r_1 >> H_1$ (Enterprise 1 is a monopoly in the area of cost reduction), then $s_1 \approx H_1$. Hence, the monopoly reveals the estimate s_1 representing, in fact, the sum of the rates of the costs of the remaining enterprises.

TASKS AND EXERCISES

2.1. Demonstrate that three properties of a planning mechanism (mentioned at the end of Section 2.1; see also Section 4.4) are valid for any Cobb-Douglas function of manufacturing costs:

$$z_i = \frac{1}{\gamma} x_i^{\gamma} r_i^{1-\gamma}, \quad i = \overline{1,n}, \quad \gamma > 1.$$

2.2. Estimate a relative increase in the total costs of a corporation under the conditions of *egalitarianism* (equal output of all enterprises).

2.3. Given the pricing mechanism $P = \sqrt{\ell c}$, find a *counter-expensive domain* (the set of net return rates meeting condition (2.19)).

2.4. Suppose the minimum cost price of a product is $c_{\min} = 100$, the limit price constitutes $l = 1000$, and the pricing mechanism is defined by

$$P = c + 0.2\ (1000 - c).$$

Evaluate the corresponding optimal price of the product.

2.5. Within the framework of the model stated in Section 2.4, estimate the level of data distortion if the hypothesis of weak impact is rejected.

Chapter 3

Incentive Mechanisms

In Chapter 1 we distinguished several types of control: staff control, structure control, institutional control, motivational control, and informational control. Motivational control has been most intensively studied to date; it represents control of preferences and goals of OS participants. Hence, we start with models of motivational control.

To begin with, consider an example of the interaction between one principal and one agent [19, 20]. The agent chooses his nonnegative action y (e.g., production output expressed in money); the goal function of the agent (see Figure 3.1) is defined by the difference between his or her income 0.5 y (the agent gains 50% income) and the "penalty" for plan nonfulfillment: $\frac{1}{4}(x - y)^2$; here x means the plan – production output desired by the principal.

Incentive-compatible mechanism is remarkable because the agent benefits from performing the plan. The following questions arise then:

1. Does the principal need exact fulfillment of the plan by the agent?
2. If yes, how could it be ensured?

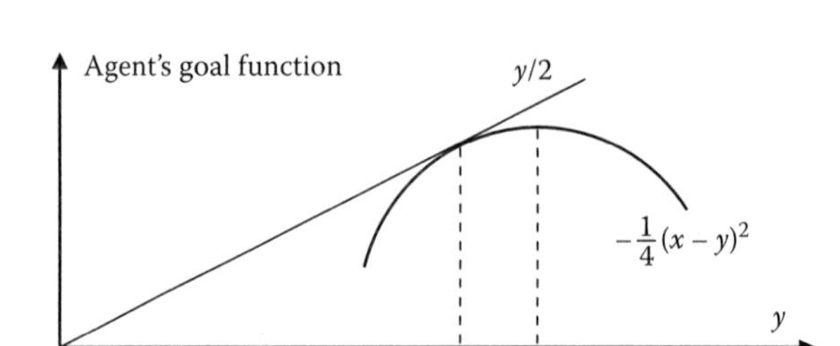

Figure 3.1 The goal function of the agent.

The first question appears not as easy as it could seem. Simple computations indicate that the agent benefits from choosing the action $(x + 1)$, that is, he or she would strive for overfulfilling the plan by the unity. Hence, if the principal requires $y = 5$ units of products (exactly this quantity, since the excess is difficult to sell), he or she should assign the plan $x = 4$. We emphasize that the plan differs from the action desired by the principal! Such a situation is common in control of active agents. The principal has to predict their behavior and assign the plan based on the forecast. The described mechanism is not incentive-compatible (the agent does not perform the plan). Nevertheless, sensus communis suggests that it would be good to have incentive-compatible mechanisms. Fortunately, there exists a wide range of the incentive/penalty functions such that an optimal plan is proven to be incentive-compatible. An example is provided in Figure 3.2.

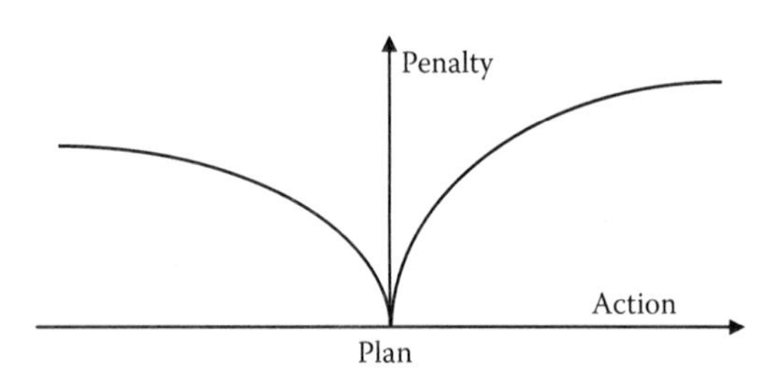

Figure 3.2 An example of the penalty function.

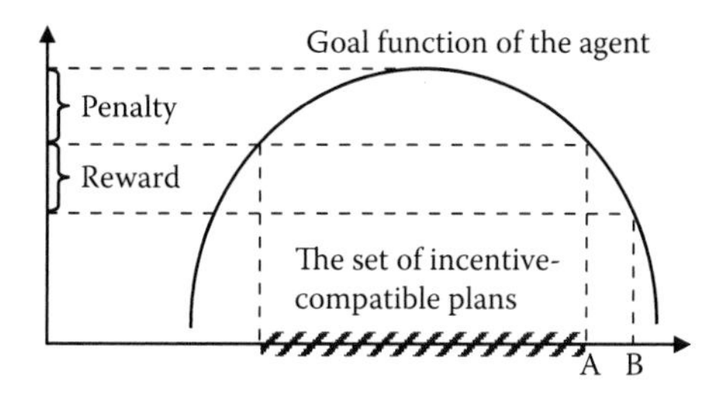

Figure 3.3 An example of the set of incentive-compatible plans.

Such functions have been shown to possess an optimal (in the view of the principal) incentive-compatible plan [20, 22]. The plan should be optimal exactly over the set of incentive-compatible plans (beneficial to the agents). Figure 3.3 demonstrates an example of the set of incentive-compatible plans, where the agent is equally penalized for the plan nonfulfillment or overfulfillment.

One easily observes that any plan belonging to the set of incentive-compatible plans is beneficial to the agent. If the principal is interested in maximal production output, the optimal plan corresponds to point A in Figure 3.3. For further increasing the output, the principal has to either enlarge the penalty or pay a reward for the plan fulfillment. In the latter case the optimal incentive-compatible plan is defined by point B in Figure 3.3.

The considered example demonstrates that coordination of interests of the principal and the agent is not a trivial problem. This fact was realized in the 1960s and 1970s, when *theory of contracts* appeared. One of the first of these was the Azariadis-Baily-Gordon (ABG) [9, 10, 60] model, which intended to explain the difference between efficient (predicted by labor economics [134, 138]) and observed wage levels—see the survey [71].

Intensive parallel development from 1970 to 1990 of the *theory of contracts* (TC) by Grossman et al. [65, 69, 112, 119]; the *theory of active systems* (TAS) by Burkov et al. [20, 22, 26, 126]; and the *theory of hierarchical games* (THG) by Germeier et al. [56, 58, 82, 86]* led to a well-developed and diversified theory of mathematical models of incentives in organizations (see the survey [26] and monographs/textbooks [17, 19, 87, 127, 146]). Basics of this theory are presented in this chapter.

3.1 Incentive Problem

The *incentive problem* gives an example of motivational control. The problem of employee motivation has been studied intensively from different points of view (incentives, stimuli, psychological, legal, and financial aspects), both in practical and theoretical perspective; nevertheless and unfortunately, today one would hardly find formal models describing the reaction of a man to a nonfinancial reward. Yet mathematical models to forecast human behavior as a corresponding reaction to the reward are still demanded by human relations specialists. On the other hand, the mathematical model of financial incentives is available and is discussed herein. It seems possible to construct models of moral stimuli in a similar way. However, when constructing a certain model of financial incentives, we make realistic hypotheses (e.g., an enterprise seeks to maximize its profit). Contrariwise, development of moral stimuli models requires speaking about different reactions of a subject to different stimuli. Such hypotheses are not trivial to reason. Current formal models of moral stimuli appear vulnerable to criticism, and psychology does not provide a researcher with a necessary base. That is why here we concentrate on the mechanisms of financial incentives.

---

* In brief, the differences between these three scientific schools consist in the following. TC analyzes incentives under stochastic uncertainty, when an agent is risk-averse (has a concave utility function). In TAS agents are typically considered risk-neutral, but more attention is paid to incentive compatibility and applications. THG focuses mostly on mathematical aspects and dynamic models.

Consider an OS composed of a single *principal* (a manager, a superior) and a single *agent* (a subordinate; a clerk or a worker) (see Figure 1.11 or Figure 1.12c).

The agent chooses the action $y \in A = \Re_1^+$, which is interpreted as working time or product output. The principal chooses a control; for the incentive problem it is a relationship between the agent's reward and his or her action (see the game Γ_2 in Section 1.3). This relationship is called an *incentive function* or an *incentive scheme.*

We will characterize the model of OS in terms of components introduced in Chapter 1 (i.e., a staff, a structure, goal functions, feasible sets, information, and sequence of moves). The staff includes the principal and the agent. The structure has two hierarchical levels. The principal chooses an incentive function, while the agent chooses an action. The set of agent's feasible actions is limited by nonnegative real values (measured in, e.g., hours, pieces, or kilograms). We will believe that the incentive function $\sigma(y)$ is nonnegative and piecewise continuous.

The principal gains a certain profit as the result of activity performed by the agent; for instance, the former sells on the market the products manufactured by the latter. The goal function of the principal is defined by the difference between the *income function* $H(y)$ and the reward $\sigma(y)$ paid to the agent:

$$\Phi(\sigma(\cdot),y) = H(y) - \sigma(y),$$

The goal function of the agent is determined by the reward received from the principal with subtraction of the corresponding costs of performing the action y:

$$f(\sigma(\cdot),y) = \sigma(y) - c(y),$$

where $c(y)$ means the *cost function* of the agent.

Suppose the income function is nonnegative for any action y and attains maximum at $y \neq 0$:

$$\forall y \geq 0 \; H(y) \geq 0, \quad 0 \notin \operatorname*{Arg\,max}_{y \geq 0} H(y).$$

As far as the cost function is concerned, suppose it is nonnegative, nondecreasing, and vanishes in the origin: $\forall\, y \geq 0$: $c(y) \geq 0$ and $c(0) = 0$. From a formal view, two last assumptions are not of crucial importance; the zero point represents a good reference. Consider the interpretation of the last condition. The agent's decision to choose the zero action means a refusal to work, which leads to "doing nothing" and thus results in the zero cost.

Let us state the following *control problem* for incentive scheme design. The agent chooses actions from the set of feasible actions ensuring maximum of his or her goal function given the relation between his or her actions and reward: $P(\sigma(\cdot)) = \operatorname{Arg}\max_{y \geq 0} [\sigma(y) - c(y)]$. This game is categorized as the game Γ_2 with side payments (see Section 1.3). The agent's response to control (the incentive function $\sigma(\cdot)$) is the set of actions ensuring maximum of his or her goal function (defined as the difference between incentives and costs). The principal can forecast agent's behavior; hence, the goal function of the principal depends on the actions of the agent and on the incentive scheme chosen. The principal evaluates the minimum of his or her goal function over the set of all actions attaining maximum to the goal function of the agent (this meets the maximin principle). Next, the principal maximizes this minimum (i.e., his or her guaranteed result) via a proper choice of the relationship between the incentives and the action of the agent: $\min_{y \in P(\sigma(\cdot))} \Phi(\sigma(\cdot), y) \to \max_{\sigma(\cdot)}$. Formally, one derives an intricate problem even for a specific form of the goal function. However, it is possible to guess the solution first and then to prove its optimality.

Assertion 3.1

Assume that a certain incentive scheme $\sigma(\cdot)$ has been used by the principal, and under this scheme the agent has chosen the action $x \in P(\sigma(\cdot))$. Let another incentive scheme $\tilde{\sigma}(\cdot)$ be involved such that it vanishes everywhere with the exception of the point x and coincides with the previous incentive scheme at the point x. Then under the

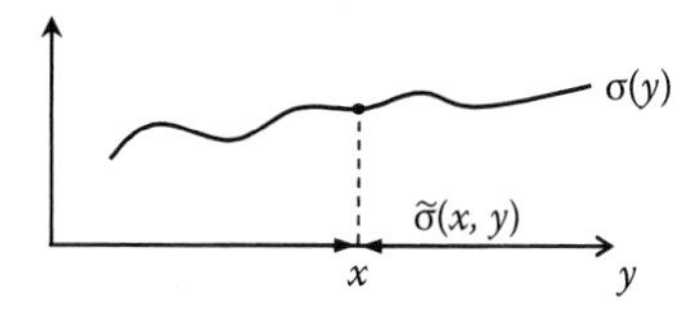

Figure 3.4 Illustrating Assertion 3.1.

new incentive scheme the same action of the agent would ensure the maximum of his goal function.

In other words, imagine a certain incentive scheme is employed by the principal and the action x is chosen by the agent; the principal says, "I'm modifying the incentive scheme—there will be no reward everywhere with the exception of the point x, and this point will be paid as before" (see Figure 3.4). In this case, the agent would again choose the "old" action $x \in P(\tilde{\sigma}(\cdot))$, with

$$\tilde{\sigma}(x, y) = \begin{cases} \sigma(x), & y = x; \\ 0, & y \neq x. \end{cases}$$

Let us provide a formal proof of Assertion 3.1. The condition that the chosen action x ensures the maximum to the agent's goal function (provided that the incentive scheme $\sigma(\cdot)$ is used) could be rewritten in the following form. The difference between compensation and costs is not smaller than in the case of choosing any other action: $\sigma(x) - c(x) \geq \sigma(y) - c(y)\ \forall y \in A$.

Now, replace the incentive scheme $\sigma(\cdot)$ for that of $\tilde{\sigma}(\cdot)$ and obtain the following. The incentive scheme $\tilde{\sigma}(\cdot)$ still equals the incentive scheme $\sigma(\cdot)$ at the point x. The right-hand side of the expression includes the incentive system $\tilde{\sigma}(\cdot)$, which vanishes if $y \neq x$:

$$\sigma(x) - c(x) \geq 0 - c(y)\ \forall y \neq x.$$

The first system of inequalities being valid implies the same for the second one; indeed, the right-hand side of the former has been weakened (if the difference between the income and costs makes a positive number, it would be surely larger than zero minus the costs). Hence, $x \in P(\tilde{\sigma}(\cdot))$. This completes the proof of Assertion 3.1. ●

Note that the incentive function $\tilde{\sigma}(x, y)$ follows Germeier's theorem (see Section 1.3 and [56, 58]), which characterizes the optimal solution of the game Γ_2 with transferable utility: an agent is penalized for the choice of any actions, except the plan x.

Let us employ Assertion 3.1 to analyze the following situation. Suppose the principal uses a certain incentive scheme with a complex relationship between the agent's reward and his or her actions. Assertion 3.1 claims that it suffices for the principal to involve a class of incentive schemes such that the incentive is nonzero at a single point. In other words, the principal can use a *quasi-compensatory* incentive scheme [126, 127] defined by

$$\sigma_K(x, y) = \begin{cases} \lambda, & y = x; \\ 0, & y \neq x; \end{cases}$$

where λ is a reward for the agent choosing the action x. The rational value of λ is explained below.

Thus, for any complex incentive scheme there exists a compensatory incentive scheme leading to the same choice of the agent (i.e., nothing changes both for the principal and the agent). However, the situation would be appreciably improved in the sense of the level of sophistication in the incentive problem as well as in the sense of agent's understanding how and why he or she is motivated.

Just imagine, the principal says that the incentive scheme represents "squared logarithmic tangent"; none of the subordinates will be able to understand such a scheme. It would be much easier for each subordinate to be told, "Let us conclude a contract—you choose this action and obtain this reward; if not, you obtain nothing." From practical considerations, such formulation appears simple and clear; but what is the underlying mathematical sense? In fact, we have reduced the problem of finding a certain function belonging to the set of all nonnegative-valued and piecewise continuous functions to the problem of finding two scalar values, notably, the action x and the reward λ (to be paid for choosing the action x). Evidently, it is easier to find two scalars than a function!

The Hypothesis of Benevolence

Consider the goal function of the principal. The agent's reward enters with negative sign, that is, the principal strives for minimizing the reward of the agent (it is desirable for an agent to perform his job for minimum possible reward). Quite the contrary, the agent would like to gain more under fixed costs.

At the same time, due to the presence of the hierarchy of subordination, decisions are first made by the principal. Therefore, the latter should think in the following way. What is the minimum reward of the agent when he agrees to perform the action? Obviously, the principal has to "operate" with the costs curve of the agent, i.e., tell him, "You choose this action and I compensate your costs. I will pay you nothing for a different action."

The compensatory incentive scheme takes the following form. The amount λ should be equal to the agent's costs (perhaps, with a certain value $\delta \geq 0$ added). In the view of the principal, the function $\sigma(x) = c(x) + \delta$ should be minimized, that is,

$$\sigma_K(x,y) = \begin{cases} \lambda, & y = x; \\ 0, & y \neq x; \end{cases}$$

The goal function of the agent is illustrated in Figure 3.5. The costs entering with negative sign are supplemented with the following incentive scheme. The principal gives the

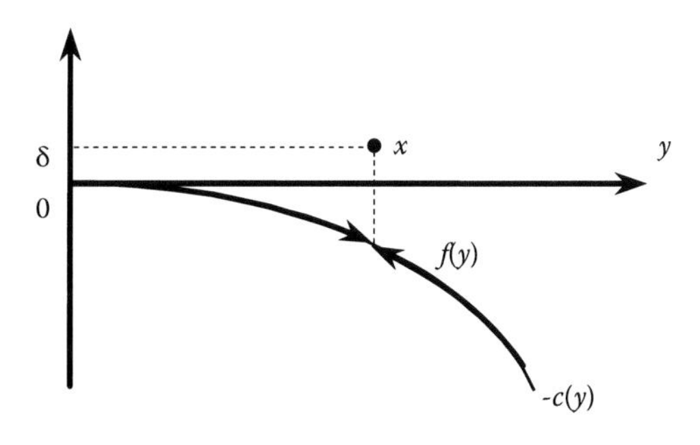

Figure 3.5 The goal function of the agent.

reward $c(x) + \delta$ at the point x, while the remaining points are described by the zero reward.

Subtracting the costs from the positive remuneration, one obtains the goal function of the agent (see the bold line in Figure 3.5). This function equals the negative costs everywhere with the exception of the point x; at this point, it constitutes δ.

Let us determine the value of $\delta \geq 0$. Note it should be minimal in the view of the principal; moreover, it depends on formulation of the problem.

For instance, suppose that the agent is benevolent to the principal; in other words, provided with two equally ranked alternatives, the agent chooses the alternative that is better for the principal. In this case, it suffices to set $\delta = 0$. Then the maximum (in fact, zero) value of the goal function of the agent (defined as the difference between the incentive and costs) is attained in two points. Notably, the first point is 0, which corresponds to no activity performed; the same zero utility is gained by the agent if he or she chooses the action x (i.e., that desired by the principal). In other cases, the goal function of the agent takes negative values. The set of goal function maxima consists of two points, and a benevolent agent will choose x (according to the *hypothesis of benevolence*).

On the other hand, assume the principal does not depend on benevolence of the agent but wishes the agent to choose a specific (nonzero) action. Then it takes only choosing δ as an arbitrarily small strictly positive number, so the value of the agent's goal function at the point x would be strictly positive. In other words, δ characterizes "the difference" between the principles of pessimism and optimism. As a matter of fact, this difference is small in this model, since the constant δ could be chosen arbitrarily small, as well.

Let us summarize the results. First, we have passed from a general incentive scheme to the one being dependent on two scalar parameters, viz. the *plan* x (what action the principal wants to gain from the agent) and the agent's reward λ. Second, we have found the value of λ being equal to the agent's costs with a constant δ added. For any problem, the described parameter δ may "contain" any "moral" component of the reward; in particular, it can be treated as a *bonus.* Formally, the agent chooses the maximum point of his goal function; however, if $\delta = 0$, the agent's goal function vanishes regardless of the action chosen (it does not matter whether the agent works at all or fulfills the plan). Evidently, such a state of things seems strange in practice-—the agent receives zero gains either working or not. Hence, the bonus δ indicates what is actually promised to the agent for his work (in the given organization), such as a *reservation wage.* Therefore, all motivational aspects lying outside the model could be incorporated into δ. What exact value should δ have is not the issue of mathematics and economics but rather of psychology and management practice.

The Principle of Costs Compensation

Suppose there exists a certain cost function of an agent, $c(y)$ (see Figure 3.6); let it be nonnegative, vanishing in the origin, and nondecreasing. The latter property signifies that the more the agent works, the higher his or her costs are. Assume that the income function $H(y)$ of a principal attains the maximum value under a nonzero action of the agent. This condition is

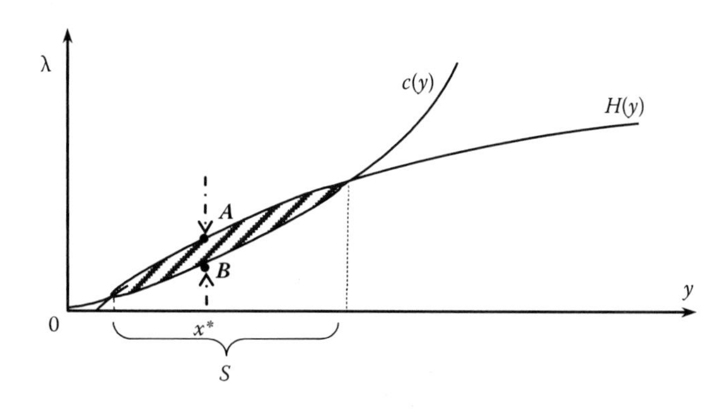

Figure 3.6 An incentive problem: the domain of compromise.

essential, since when the maximal income of the principal is ensured by the zero action of the agent, in fact, no incentive problem arises.

Now, analyze this situation both from the positions of the principal and the agent. Every point (y, λ) on the plane of Figure 3.6 represents a contract, where the principal pays λ if the agent chooses the action y, and pays zero otherwise. The zero point is characterized by the fact that, with the agent doing nothing, his or her costs are zero. If the principal pays nothing to the agent for being inactive, the latter obtains zero utility. Hence, the lower-bound estimate of the agent's payoff makes 0 (zero action leads to zero gain). Consequently, the agent would agree to perform a certain action if his or her reward (given by the principal) turns out not smaller than the cost of the agent. Therefore, the following constraint is immediate: the reward should not be smaller than the agent's costs. So, the agent is "satisfied" with all points (contracts) located above the cost function $c(y)$ (see Figure 3.6).

Now, let us focus on the principal. According to the principal's view, he or she obtains a certain level of utility in the case of the zero action of the agent (i.e., if the principal pays nothing to the agent). Definitely, he would not pay the agent more than the income resulting from the agent's activity. That is, in the view of the principal, only those contracts (i.e., combinations of actions and rewards) seem feasible that lie below the income function $H(y)$ of the principal (see Figure 3.6).

Remember a principal seeks to minimize payments to an agent provided that the latter chooses the required action. This means that the point of optimal contract (under the hypothesis of benevolence) should be located on the lower boundary of the shaded domain in Figure 3.6. In other words, *the reward should be exactly equal to the costs of the agent.* This important conclusion is referred to as the *principle of costs compensation* [126, 127]. It claims that the principal has to compensate merely the agent's costs for motivating the agent to choose a specific action.

Intersection of the two domains (that of payments being greater than the agent's costs and the ones smaller than the principal's income) defines a certain domain. First, note that $S = \{y \in A | H(y) \geq c(y)\}$ is the *set of feasible actions*, which includes agent's actions such that the income (as the "result" of his activity) does not exceed his costs. The totality of the action set S and rewards paid for the actions, which satisfy both the principal and agent, is called the *domain of compromise*, or *negotiation set*. The domain of compromise is shaded in Figure 3.6.

The Decomposition Principle and the Aggregation Principle

We have studied the elementary system, which consists of a single principal and a single agent. To proceed, let us complicate the setting. Consider a system composed of several agents subordinated to a single principal. That is, we focus on an elementary fan-type structure (Figure 3.7 and Figure 1.12d) instead of the one given in Figure 1.12c.

Suppose the costs of every agent depend on his or her actions and on the actions of the remaining agents, as well. This means that the reward would be a certain function of the actions performed by all agents.

The principal chooses a *plan*, understood as a vector of agents' actions desirable for the principal, and an agent is paid depending on the chosen action. Evidently, nothing should be paid if an agent chooses an action differing from the corresponding component of the plan. How much should an agent

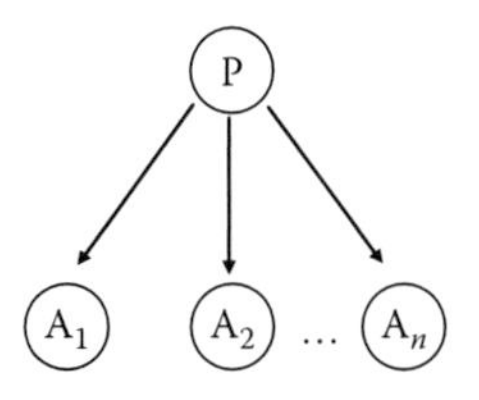

Figure 3.7 A fan-type structure.

be paid provided that he or she chooses the planned action? The agent should be paid "something close to" his or her costs; however, the costs of every agent depend on the actions of all agents. Keep in mind that the payment should be great enough to make the planned action beneficial to the agent. As it has turned out, if an agent performs the required action, his or her costs should be compensated irrespective of the actions chosen by the remaining agents. This rule is called the *decomposition principle* (see Section 3.3 and [128]).

Let us study the situation when the principal appears unable to observe the action of every agent; it is only possible to observe certain aggregated data, that is, the *outcome* of collective activity. What incentive scheme should then be used? It has been found that, if for every outcome the principal can calculate the minimum costs of all agents implementing this outcome (agents choose a profile of actions that lead to the desired outcome under minimum total costs), then the efficient incentive scheme is defined as follows. The minimum costs are compensated for each agent provided that the outcome of collective activity meets the requirements of the principal. Moreover, the principal loses nothing as opposed to the case when he or she observes all individual actions of agents. Notably, designing the efficient incentive scheme does not require observing individual actions of every agent; it suffices to know the result of their collective activity and to compute minimum costs of the agents (needed to achieve the outcome). This is the *aggregation principle* [129].

3.2 Basic Incentive Mechanisms

Let us discuss the *basic incentive schemes* (*mechanisms*) [126, 127] in single-agent deterministic organizational systems, that is, the systems operating under complete information on all essential (internal and external) parameters. A *compensatory* (*C-type*) *incentive scheme* is the optimal basic incentive scheme.

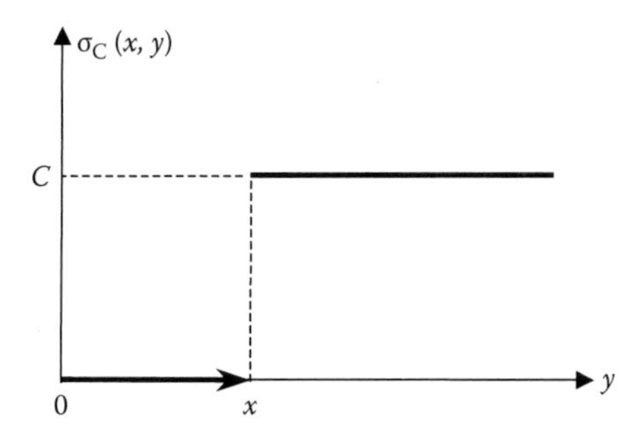

Figure 3.8 A jump incentive scheme.

A *jump (J-type) incentive scheme* is characterized by the following feature. The agent obtains a fixed reward C provided that his action appears not smaller than a planned action x; otherwise, the agent has zero reward (see Figure 3.8):

$$\sigma_C(x, y) = \begin{cases} C, & y \geq x; \\ 0, & y < x. \end{cases} \tag{3.1}$$

J-type incentive schemes may be treated as a *lump sum* payment that corresponds to the reward C under a given result (e.g., an output level being not smaller than a predetermined threshold or number of working hours). Another interpretation is when the agent is paid for hours worked; for instance, the reward then corresponds to a fixed wage under full-time occupation.

Proportional (*linear*) (*L-type*) *incentive schemes,* or fixed reward rates, are widely used in practice. For instance, a fixed per-hour rate implies the same wage is paid for every hour worked, while piece-based wage implies a fixed reward for every unit of the manufactured product. In both schemes the agent's reward is proportional to his action (e.g., hours worked and product units manufactured) and the wage rate $\lambda \geq 0$ represents the proportionality coefficient (see Figure 3.9):

$$\sigma_L(y) = \lambda\, y. \tag{3.2}$$

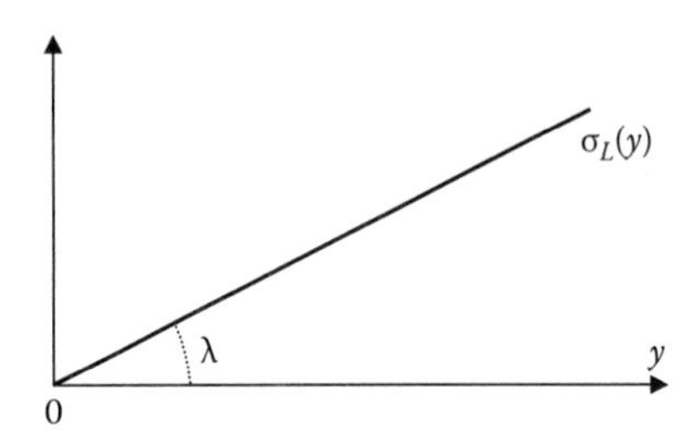

Figure 3.9 A proportional (linear) incentive scheme.

Suppose that a linear incentive scheme is employed and the cost function of the agent is continuously differentiable, monotonous, and convex. Optimal action y^* of the agent (maximizing his goal function) is defined by the formula $y^* = (c')^{-1}(\lambda)$, where $(c')^{-1}(\cdot)$ stands for the inverse derivative of the agent's cost function. Note that the principal more than compensates the agent's cost by choosing the action y^*; actually, the principal overpays the following amount: $y^* c'(y^*) - c(y^*)$. For instance, suppose the agent has the income function $H(y) = by$, $b > 0$, while his cost function is convex: $c(y) = ay^2$, $a > 0$. In this case, for any feasible action of the agent the principal pays twice compared to the optimal payment.

Therefore, under a convex cost function of the agent, efficiency of the proportional scheme is not greater than that of the compensatory one. A curve of the agent's goal function (under a proportional incentive scheme used by the principal) is demonstrated by Figure 3.10.

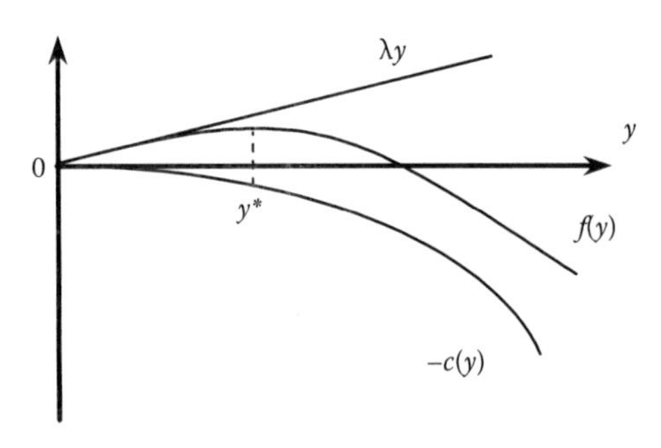

Figure 3.10 A goal function of the agent: the principal uses an L-type incentive scheme.

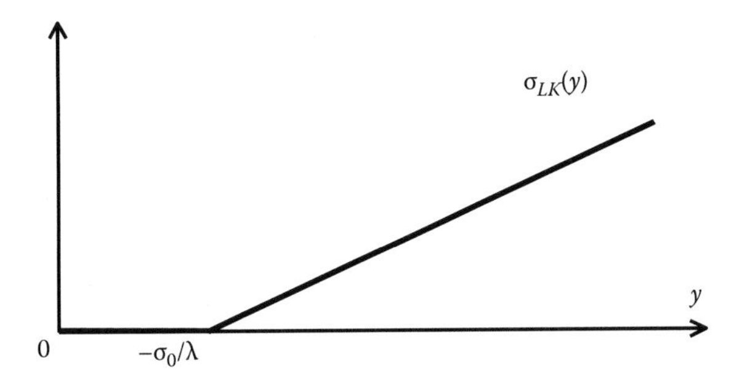

Figure 3.11 A "linear" incentive function.

Low efficiency of proportional incentive schemes described by the formula $\sigma_L(y) = \lambda y$ is subject to non-negativity of rewards. Assume that the reward may be negative for some actions (note that these actions are probably never chosen, as shown in Figure 3.11), that is, $\sigma_{LK}(y) = \sigma_0 + \lambda\, y$ with $\sigma_0 \leq 0$. Then, under a convex cost function of the agent, efficiency of the proportional incentive scheme $\sigma_{LK}(\cdot)$ could equal that of the optimal (compensatory) incentive scheme.

It suffices to involve the following expressions to substantiate the previous assertion (also see Figure 3.12):

$$x^*(\lambda) = c'^{\,-1}(\lambda), \quad \sigma_0(\lambda) = c(c'^{\,-1}(\lambda)) - \lambda\, c'^{\,-1}(\lambda).$$

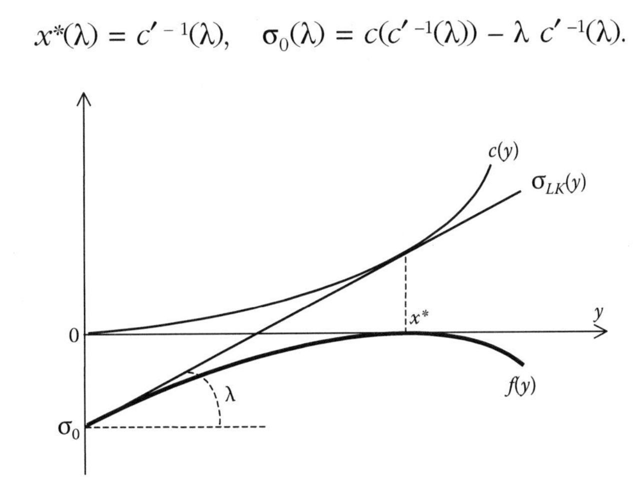

Figure 3.12 A goal function of the agent: the principal uses the incentive scheme $\sigma_{LK}(\cdot)$.

The optimal value λ^* of the wage rate is chosen from the maximum condition for the goal function of the principal:

$$\lambda^* = \operatorname{Arg}\max_{\lambda \geq 0} \, [H(x^*(\lambda)) - \sigma_{LK}(x^*(\lambda))].$$

Incentive schemes based on income distribution (D-type) employ the following idea. Since the principal represents preferences of the whole system, principal's income can be equated to that of the whole organizational system. Hence, one may base an incentive scheme of the agent on the income obtained by the principal; in other words, one may set the agent's reward equal to a certain (e.g., fixed) share $\gamma \in [0; 1]$ of the principal's income:

$$\sigma_D(y) = \gamma \, H(y). \tag{3.3}$$

We underline that C-, J-, L-, and D-type incentive schemes are parametric. Notably, it suffices to choose the pair (x, C) for specifying the jump incentive scheme. Defining the proportional incentive scheme requires designating the wage rate λ. Finally, one should only select the income share γ to describe the incentive scheme based on income distribution.

The aforementioned incentive schemes are elementary and serve as blocks of a "kit"; using these blocks, it is possible to construct complex incentive schemes (referred to as *derived schemes* with respect to the basic ones). Thus, we should define operations over the basic incentive schemes for making such a "construction" feasible. Dealing with a single-agent deterministic OS, the researcher may confine himself to the following three types of operations.

The first-type operation is a transition to a corresponding incentive *quasi-scheme*, that is, the reward is assumed to be zero everywhere, with the exception of the planned action. In the complete information framework, "nulling" the incentive in all points (except the plan) does not modify the properties of the incentive scheme under the hypothesis of benevolence. Therefore, in the sequel we will not dwell on differences between a specific incentive scheme and its counterpart (a scheme derived from the initial scheme by means of the first-type operation).

The second-type operation is the *composition,* that is, employment of different basic incentive schemes in different subsets of the set of feasible actions. The resulting incentive systems are called *composite*; an example is provided by an LL-type incentive system, where small actions are compensated using a lower rate, while bigger actions are compensated using a higher rate.

The third-type operation is represented by algebraic addition of two incentive schemes (this is possible as the reward enters the goal functions additively). The result of such an operation is referred to as a *cumulative incentive scheme.*

For instance, Figure 3.13 shows an incentive scheme of J+L-type (a tariff plus-bonus incentive scheme), derived by summing-up jump and linear incentive schemes. Thus, the *basic incentive schemes* include the ones of J-, C-, L-, and D-type, as well as

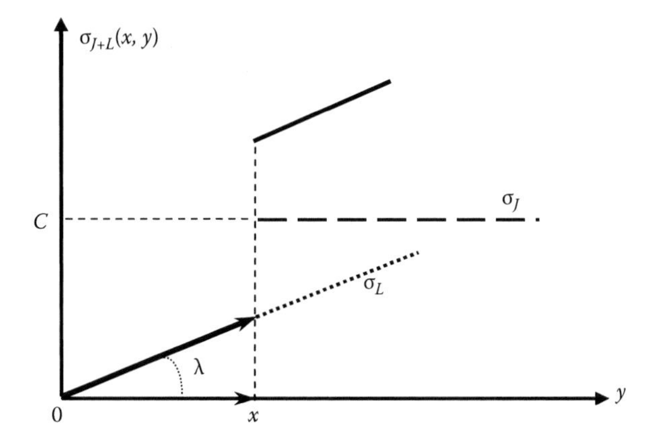

Figure 3.13 An incentive scheme of J+L-type (a cumulative incentive scheme).

any schemes derived from them through the above-mentioned operations.

It is known that the incentive schemes derived for the basic incentive schemes discussed cover all personal wage systems used in practice [3, 127].

Incentive Mechanisms in a Single-Agent System

Consider the goal function of the principal. In the case of a compensatory incentive scheme with $\lambda = c(x) + \delta$, it represents the principal's income with the deducted reward of the agent: $\{H(x) - c(x) - \delta\}$. If the reward of the agent coincides with his costs, the payoff of the principal is the difference between the income and costs. Hence, it is necessary to choose a certain x^* attaining maximum to the difference $\{H(x) - c(x)\}$ over all $x \in S$.

Recall that we started from a complex incentive scheme and simplified the setting to an incentive scheme with two parameters; subsequently, we evaluated the first parameter, λ. There remains one thing to do: to find the second parameter (viz. the plan x^*). The latter must maximize the difference between the principal's income and the payment (exactly matching the agent's costs). As the result, an optimal solution to the incentive problem would be provided by a compensatory scheme with the following property. The reward coincides with the agent's costs, while the *optimal plan* is equal to the one maximizing the difference between the principal's income and the agent's costs. Here is the final formula for the optimal solution:

$$x^* \in \operatorname*{Arg\,max}_{x \in S} \{H(x) - c(x)\}.$$

Let us analyze the derived formula for the optimal plan x^*. It means the optimal plan maximizes the difference between the principal's income and the agent's costs (this amount is equal to "the thickness" of the domain of compromise; see Figure 3.6). Consider the point x^*; the slope of the tangent to the principal's income function equals that of the tangent to the agent's costs function. In economics, this point represents an optimal one where the *marginal output* equals the *marginal costs.*

Therefore, the point x^* appears optimal for the principal; it is denoted by *B* in Figure 3.6. However, study a situation when the agent moves first. The agent suggests to the principal, "I will do this action and you will pay this reward to me." Then the principal accepts or rejects this offer.

An immediate question is, "What should the agent suggest?" The agent should offer to the principal the same action x^* and ask for the reward described by point A in Figure 3.6. Then all "profit" $[H(x^*) - c(x^*)]$ will be received by the agent. In other words, in this game the leader—the person making the *first move*—actually wins [27, 127]. If the principal is the leader, he or she "strands" the subordinate; otherwise, the latter "strands" the former. Both would agree with the opponent's offer within the framework of the formal model.

To proceed, we study the following situation. Suppose there exist goal functions of the principal and the agent that include income of the principal and costs of the agent, respectively. The variable parameter (i.e., an incentive function) is an internal characteristic of the system, describing interaction between the principal and the agent (the latter obtains exactly what the former has provided). Sum up the goal functions of the principal and agent; the values of the incentive function are reduced and the difference between the income and costs remains the same. Hence, the action x^* solving the incentive problem maximizes the sum of the goal functions; that is, this action of the agent is Pareto optimal.

Every point of segment AB in Figure 3.6 corresponds to some scheme of profit distribution. We have examined two extreme cases:

1. The principal gets all profit $[H(x^*) - c(x^*)]$.
2. The agent gets all profit $[H(x^*) - c(x^*)]$.

It is possible to define a *compromise* between them. The principal and agent can negotiate a certain distribution of the profit (e.g., half and half). Then the agent obtains a share of the profit in addition to the compensated costs. An alternative rule is that of a fixed norm of profitability. A similar technique is used to analyze numerous modifications of incentive problems.

We have found the solution to the basic incentive problem; it is given by the compensatory incentive scheme with the plan x^*. But is the solution unique? In fact, no, and the reasoning is quite simple. We know the optimal (quasi-compensatory) incentive scheme motivating the agent to choose x^*, while the principal pays exactly the amount of the agent's costs.

Consider other incentive schemes motivating the agent to choose the same action, and the principal to pay the same reward. All such incentive schemes must pass through point $(x^*, c(x^*))$.

Assertion 3.2

The agent chooses the action x^*, if the incentive function passes through point $(x^*, c(x^*))$ and is not greater than the agent's costs at the remaining points.

We leave the proof up to the reader (see [127]).

Select any incentive scheme from the ones shown in Figure 3.14; again, it would motivate the agent to choose the same action while the principal pays the same reward.

It is also possible to use a jump incentive scheme; that is, no reward is paid for the actions being smaller than the plan, while fulfilling the plan gives the reward being not smaller than the costs (a *lump sum payment*). One may take a monotonous incentive scheme that passes through point $(x^*, c(x^*))$ and is lower than the cost curve at the remaining points. In other words, any function passing through the point $(x^*, c(x^*))$ and lying lower than the cost function would be a solution to the incentive problem.

Table 3.1 [127] shows comparative efficiency for different combinations of the basic incentive schemes.

Note here we proceed from the assumption that the agent's cost function is convex and monotonous. In Table 3.1,

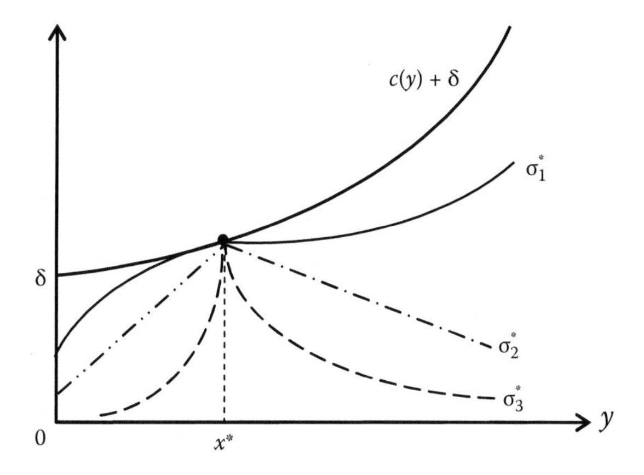

Figure 3.14 Optimal incentive schemes.

efficiency of seven basic incentive schemes is compared in the following way. The sign "≥" in a cell indicates that efficiency of the incentive scheme in the corresponding row is not smaller than efficiency of the one located in the corresponding column. Other signs should be interpreted by analogy; the symbol "?" means that comparative efficiency of the incentive schemes having L- and D-type depends on the agent's cost function and the principal's income function in each specific situation.

Table 3.1 Comparative Efficiency for Different Combinations of the Basic Incentive Schemes

	C	*J*	*L*	*LC*	*D*	*L+J*	*LL*
C	=	=	≥	=	≥	=	=
J	=	=	≥	=	≥	=	=
L	≤	≤	=	≤	?	≤	≤
LC	=	=	≥	=	≥	=	=
D	≤	≤	?	≤	=	≤	≤
L+J	=	=	≥	=	≥	=	=
LL	=	=	≥	=	≥	=	=

Parametric Representation of Goal Functions

Up to this subsection, we have studied incentive problems without constraints imposed on the class of goal functions of the agents (to be more precise, on the incentive functions). A common situation in practice, the class of goal functions of the agents is given in a parametric form $f(x, y)$, where $x \in X$ and X specifies the parameter set for x. Rewrite $f(x, y)$ in the following form:

$$f(x, y) = h(y) - \chi(x, y),$$

with $h(y) = f(y, y)$, $\chi(x, y) = h(y) - f(x, y)$.

Naturally, the parameter x should be interpreted as a *plan* expected from the agent (i.e., an action of the agent desired by the principal). On the other hand, $\chi(x, y)$ should be viewed as a *penalty* applied in the case of a deviation of the actual action from the plan ($\chi(x, y) \geq 0$, $\chi(x, x) = 0$). As a result, the incentive problem becomes, in fact, a planning problem under complete information. Moreover, the optimal planning problem is reduced to the game Γ_1 (see Section 1.3). Definition of the game solution is referred to as the *principle of optimal planning with the action forecast.* The set

$$S_0 = \{x | \max_y f(x, y) = h(x)\}$$

is called the *set of incentive-compatible plans.* Evaluating the optimal plan over the set of incentive-compatible plans is known as the *principle of optimal incentive-compatible planning.*

The following question is immediate. When is the principle of optimal planning with the action forecast equivalent to the principle of optimal incentive-compatible planning? In other words, when is the optimal plan fulfilled? Probably, the most famous and elegant form of the sufficient conditions of incentive compatibility is provided by the so-called *triangle inequality* for the penalty function [22]:

$$\forall\ x, y, z, \quad \chi(x, y) \leq \chi(x, z) + \chi(z, y).$$

3.3 Incentive Mechanisms in Multi-Agent Systems

In Sections 3.1–3.2 we investigated an organizational system composed of a single principal and a single agent. Focusing on this elementary model, we solved the incentive problem with the goal function of the principal being represented by the difference between the income and the reward paid to the agent. In addition, it has been shown that the following compensatory incentive scheme turns out optimal: the agent obtains a reward coinciding with his costs when the plan is fulfilled or zero reward otherwise. The optimal plan has been defined as the one maximizing the difference between the principal's income and the agent's costs.

Now let us endeavor to solve the same incentive problem (with the same practical interpretation) in more complicated cases. For instance, consider a system with several agents subordinated to a single principal.

We emphasized in Chapter 1 that any organizational system (more specifically, its model) is described by the following parameters: a staff, a structure, goal functions, feasible sets, and information. For the system under consideration, the staff obviously includes the principal and n agents. All agents are at a lower level of the hierarchy, and the principal is at the upper one. The goal functions and feasible sets are defined by

$$y_i \in A_i, \quad \sigma_i(y_i), \quad i \in N = \{1, 2, \ldots, n\}.$$

Assume that agent i chooses an action y_i from the set A_i, while the principal chooses the reward $\sigma_i(y_i)$ of agent i, which depends solely on the action chosen by this agent.

The goal function of the principal represents the difference between his income $H(y)$ resulting from agents' activity (note $y = (y_1, y_2, \ldots, y_n) \in A' = \prod_{i \in N} A_i$ stands for the vector of agents' actions) and the total rewards paid to the agents:

$$\Phi(\sigma(\cdot), y) = H(y) - \sum_{i \in N} \sigma_i(y_i).$$

Thus, we have extended the previous (basic) model—the agent's goal function has the same form but includes the subscript i. It should be emphasized that there exist n goal functions of this type; that is, agent i receives the reward for his or her actions from the principal and incurs costs depending only on his or her actions:

$$f_i(\sigma(\cdot), y_i) = \sigma_i(y_i) - c_i(y_i),\ i \in N.$$

Compare the goal function in the previous model with the one used for the fan structure with several agents. The payment to agent i depends only on his or her actions (and so do the costs), while the goal function of agent i depends only on his reward and actions; in other words, the agents are de facto independent from each other. So a full-fledged game among the agents does not take place, as the payoff of every agent depends only on his actions (it does not matter what the remaining agents do).

Thus, the considered system may be decomposed into n independent single-agent–single-principal subsystems studied separately (see Figure 1.11). Assertions 3.1–3.2 are then applied to each of them.

Analysis of a single-agent model implies that every agent could be motivated independently; moreover, for every agent it suffices to compensate his or her costs. Therefore, the problem under consideration could be solved by the following technique. We know that the principal's income makes $H(y)$

and that he or she should pay $c_i(y_i)$ to agent i for choosing the action y_i. Let us substitute the optimal incentive scheme into the principal's goal function to derive the difference $H(y) - \sum_{i \in N} c_i(y_i)$. Next, find a plan x maximizing the principal's goal function over the set of feasible vectors of agents' actions:

$$x = \operatorname{Arg}\max_{y \in A'} \left[H(y) - \sum_{i \in N} c_i(y_i) \right].$$

This optimization problem is easily solved in most cases.

Let us examine the obtained result once again. What would be the decision of a specific agent? The agent's goal function depends only on his or her action; under a known incentive scheme (reported to the agent by the principal), the agent would find his or her action maximizing his or her goal function (notably, the reward minus the costs). Since the agent's goal function depends only on the action, the latter would be independent of what the remaining agents do. In the stated sense, the agents are independent—that is, every agent possesses a dominant strategy. So, in the considered model an increase in the number of agents does not enlarge the complexity of interactions—as before, the interaction between the principal and agents can be considered independently. Therefore, we further complicate the model.

The first step introduces a certain constraint on the total *wages fund* (otherwise the agents are not related at all). Such organizational systems are referred to as the *systems with weakly related agents*. Thus, we add the budget constraint $\sum_{i \in N} \sigma_i(y_i) \le R$. In other words, the total reward of all agents should not exceed a given threshold, R.

Now, analyze the consequences. The behavior of the agents remains unchanged, since the agent's goal function depends only on his or her actions. On the contrary, the problem of the principal is changed. The principal knows that using the optimal incentive scheme he or she should compensate the costs of every agent. However, the principal has an extra constraint—he or she should maximize not over all combinations of agents' actions but only over the actions satisfying the budget constraint. The problem has been changed: it is necessary to maximize over the set A' intersected with the set of the action vectors y such that $\sum_{i \in N} c_i(y_i) \leq R$; this means the budget constraint is met.

In the principal's view, the optimal solution is still to compensate the costs of every agent, when the latter fulfills the plan, that is, the structure of the incentive scheme remains the same. As before, the agent's goal function depends only on the incentive scheme (established by the principal) and on the action of the mentioned agent. And the agent does not care about the budget constraint, making his choice under the incentive scheme reported. Thus, we have derived a constrained optimization problem:

$$x = \operatorname{Arg} \max_{\left\{ y \in A' \mid \sum_{i \in N} c_i(y_i) \leq R \right\}} \left[H(y) - \sum_{i \in N} c_i(y_i) \right].$$

Hence, the incentive problem is solved by reducing to the constrained optimization problem. Consider the following example.

Example 3.1 [127]

There are two agents ($n = 2$), and the principal's income function is defined as the sum of agent actions: $H(y) = y_1 + y_2$. Assume that the cost function of agent i is quadratic:

$$c_i(y_i) = \frac{y_i^2}{2r_i}, \quad i = 1, 2.$$

The constant $r_i > 0$ could be interpreted as efficiency of the agent's activity or agent's skill level (the higher is the level, the lower are the costs).

Under a compensatory incentive scheme, the principal's goal function is represented by the difference between the total action of the agents and their total costs. The goal function should be maximized over $y_1 \geq 0$ and $y_2 \geq 0$ provided that the total costs compensated do not exceed the wages fund R:

$$\begin{cases} y_1 + y_2 - \dfrac{y_1^2}{2r_1} - \dfrac{y_2^2}{2r_2} \to \max\limits_{(y_1, y_2) \geq 0}; \\ \dfrac{y_1^2}{2r_1} + \dfrac{y_2^2}{2r_2} \leq R. \end{cases}$$

The incentive problem is reduced to the evaluation of two parameters: y_1 and y_2. There are several regimes: in the first regime the budget constraint is not binding and y_1 and y_2 are found from the first-order conditions as

$$y_1^{\max} = r_1,\ y_2^{\max} = r_2,$$

$$\frac{(y_1^{\max})^2}{2r_1} + \frac{(y_2^{\max})^2}{2r_2} \leq R.$$

These formulas express an unconstrained maximum; indeed, differentiating with respect to y_1, one obtains $1 - y_1/r_1$. The costs of agent 1 constitute $r_1/2$. Hence, with $(r_1 + r_2)/2 \leq R$, the optimal solution is provided by $x_1 = r_1$, $x_2 = r_2$. If $(r_1 + r_2)/2 > R$, then the budget constraint is binding and the solution is obtained with Lagrange's method of multipliers. We write down the Lagrange function:

$$(y_1 + y_2) - (1 + \mu)\left(\frac{y_1^2}{2r_1} + \frac{y_2^2}{2r_2}\right) - \mu R.$$

Differentiating it with respect to y_1, we have $1 - (1 + \mu)\frac{y_1}{r_1}$. Next, equate it to zero to find the optimal action depending on the Lagrange multiplier. As the result, one obtains $y_1 = (1 + \mu) r_1$; similarly, $y_2 = (1 + \mu) r_2$. Substitute the evaluated actions into the budget constraint (it should be satisfied as equality):

$$\frac{(1 + \mu)^2 r_1^2}{2r_1} + \frac{(1 + \mu)^2 r_2^2}{2r_2} = R.$$

This implies that

$$1+\mu = \sqrt{\frac{2R}{r_1 + r_2}}.$$

Consequently, the optimal solution is defined by

$$x_i = r_i\sqrt{\frac{2R}{r_1 + r_2}} \qquad i = 1, 2.$$

Therefore, if the wages fund is smaller than the half-sum of r_1 and r_2, the optimal plans are $x_1 = r_1$, $x_2 = r_2$; otherwise, the optimal plans are given by

$$x_i = r_i\sqrt{\frac{2R}{r_1 + r_2}} \qquad i = 1, 2.$$

Note that the resulting solution is continuous; that is, for $R = (r_1 + r_2)/2$ the plans coincide. •

Note the following as well. Analyzing the incentive problem for weakly related agents, we have consumed most of the time for solving the incentive-compatible planning problem (the problem of constrained optimization that "bears no relation" to control). This is the case, since a priori we have employed the result that in an optimal compensatory incentive function the reward coincides with the costs (and the agents fulfill their plans).

Let us add another piece of complexity to the model of the incentive problem. Before we used the following argumentation. Starting from a single-agent system, proceed to the one with independent agents having no constraints, and then impose a constraint on the wages fund. Now, consider the case of *strongly related* agents; this means the following. Suppose the costs of every agent depend not only on his or her actions but also on the actions of the others. Naturally, the reward of every agent will be dependent on the actions of all agents.

The goal function of the principal is defined as

$$\Phi(\sigma(\cdot), y) = H(y) - \sum_{i \in N} \sigma_i(y),$$

where $y = (y_1, ..., y_n)$, and the goal functions of the agents are

$$f_i(\sigma(\cdot), y) = \sigma_i(y) - c_i(y),\ i \in N.$$

Two new features have been added to this model. First, the lower-level agents are highly interconnected now, so the action chosen by every single agent influences the costs of all agents. Second, we allowed the reward of an agent to depend on the whole vector of agent actions. These assumptions appreciably complicate the matter since we cannot use directly the technique developed for the single-agent model.

So how will the agents make their decisions? As before, the principal moves first, announcing an incentive scheme to the agents (i.e., each agent is informed on a relationship between the reward and action vector of all agents). The agents, being aware of the incentive scheme, should choose their actions. As the payoff of each agent depends on the actions of all agents, a certain game takes place. The outcome of this game is given by a sort of equilibrium concept, for example, that of Nash. Denote by $\sigma(\cdot) = (\sigma_1(\cdot), \ldots, \sigma_n(\cdot))$ the vector-function of rewards and write down the equilibrium formula in the agents' game (given the incentive scheme used by the principal):

$$E_N(\sigma(\cdot)) = \left\{ y^N \in A' \left| \begin{array}{l} \sigma_i(y^N) - c_i(y^N) \geq \sigma_i(y_i, y_{-i}^N) - c_i(y_i, y_{-i}^N) \\ \forall i \in N, \quad \forall y_i \in A_i \end{array} \right. \right\}.$$

Next, state the control problem:

$$\min_{y \in E_N(\sigma(\cdot))} \left[H(y) - \sum_{i \in N} \sigma_i(y) \right] \to \max_{\sigma(\cdot)}.$$

The goal function of the principal depends on the incentive scheme and on the actions of the agents. Under a fixed incentive function, the agents choose Nash equilibrium actions. Apply the principle of maximal guaranteed result to the principal's goal function on the Nash equilibria set for the agents' game (under a given incentive scheme). The resulting structure depends on the incentive function only. Then, one should maximize it by a proper choice of the incentive vector-function; that is, the principal should find a certain set of the agents' incentives maximizing the guaranteed value of his goal function over the Nash equilibria set for the agents' game.

The stated problem has almost the same form as for a single-agent system; in contrast to the multi-agent case, the problem in a single-agent system includes no sum but employs the set of maximal values of the agent's goal function. In a multi-agent system this set is replaced by the Nash equilibria set, and the total reward of the agents appears. The problem is nontrivial; first, we should evaluate the minimum of a certain functional over the set being dependent on the vector-function (which enters the functional in question) and then minimize it by a proper choice of the vector-function.

Analyze the formula defining the Nash equilibria set; apparently, this set depends on the vector-function and is described by an infinite system of inequalities. Solving any complicated problem requires guessing the solution. Such a "multi-agent" incentive problem (under external stochastic uncertainty) was formulated and explored in [78, 95, 112]. It took much time to guess the solution to the posed problem even in the deterministic case (see [128]). The underlying idea of the approach used is, in fact, very simple. Imagine that the single-agent problem includes a compensatory incentive scheme (being easy and clear). What compensatory incentive scheme should be invented to solve the multi-agent problem?

The agent's costs depend on the actions performed by the agent in question, as well as on the actions of the remaining agents. The principal declares, "Fulfill the plan and I promise to compensate the actual costs incurred regardless of the actions chosen by the other agents." This is the principle of decomposition discussed above. In other words, the incentive function takes the form

$$\sigma_i(x, y) = \begin{cases} c_i(x_i, y_{-i}), & y_i = x_i; \\ 0, & y_i \neq x_i \end{cases}, \quad i \in N.$$

Let us show that the compensatory incentive scheme presented makes fulfilling the plan the Nash equilibrium. For this, substitute the incentive scheme in the previous formula of Nash equilibrium and prove that the vector x is a Nash equilibrium. The plan being fulfilled, the corresponding costs of agent i are compensated, and the goal function of the agent makes zero. Failing to fulfill the plan, the agent has zero reward and his or her costs constitute

$$c_i(x_i, x_{-i}) - c_i(x_i, x_{-i}) \geq 0 - c_i(y_i, x_{-i}), \quad y_i \neq x_i.$$

We have derived the expression stating that "the costs with negative sign are below zero." This inequality takes place for any opponents' action profile; in other words, every agent benefits from fulfilling the plan irrespective of actions of the others. Recall that a dominant strategy of an agent is a certain action ensuring the maximum of his goal function (regardless of the actions chosen by the remaining agents). In the present case, fulfilling the plan would maximize the goal function of the agent irrespective of actions of the others. This means that plan fulfillment is a dominant strategy equilibrium (DSE) [128].

Thus, we have demonstrated that the suggested compensatory incentive scheme implements the given plan vector as a dominant strategy equilibrium of the agents' game. Is it possible to make another action profile an equilibrium and pay the agents less than their total costs? The matter is that the principal's goal function depends on the total incentives with negative sign and it would be desirable to minimize the sum.

As we have supposed in the beginning, the principal is unable to apply the penalties (the severest "penalty" is zero reward). Is it possible to motivate the agents by a nonnegative incentive function to choose a specific action vector and pay the agents less than their total costs? Let us prove that the answer is no.

Introduce a natural assumption that the agent's costs are zero (in the case of zero action chosen by the agent) regardless of what the others do:

$$\forall y_i \in A_i, \quad c_i(0, y_{-i}) = 0, \quad i \in N.$$

The goal function of every agent is given by the difference between the reward and costs. Fix a certain action vector (the principal seeks that the agents choose the actions included in the vector). Suppose that the total payment to the agents implementing a certain vector of actions is smaller than the total costs of the agents. Hence, at least one agent receives a reward less than the costs; this contradicts the assumption that an agent can ensure the zero cost by choosing the zero action given that the cost function is nonnegative.

Consequently, in addition to the fact that the compensatory incentive scheme implements the planned action vector as a DSE for the game of the agents, the principal pays the minimum reward possible. This means that this incentive scheme is optimal. The remaining issue is evaluation of the required planned action vector. Similarly to a single-agent model, one should first substitute the agents' costs into the principal's goal function (instead of the agents' incentives), and then minimize the resulting expression by a proper choice of the plan:

$$x = \operatorname*{Arg\,max}_{y \in A'} \left[H(y) - \sum_{i \in N} c_i(y) \right].$$

In other words, it is necessary to find feasible actions maximizing the principal's profit; the evaluated actions should be assigned as the plan through substitution into the compensatory incentive scheme. The problem is solved!

It should be emphasized that (by analogy to a single-agent model and a system with weakly related agents) first we have proved optimality of a compensatory incentive scheme and then solved the planning problem. In the multi-agent case, demonstrating the optimality of the decomposing incentive scheme has turned out to be intricate (in contrast to a single-agent system), as we need to analyze the agents' game. However, we have successfully guessed the solution and "decomposed" the game; that is, the principal has decomposed interaction of the agents by means of control. Applying such control actions, which decompose interaction of the agents, reduces their game to a game admitting a DSE, which is referred to as the *decomposition principle of the agents' game* [128].

3.4 Mechanisms of Distributed Control

We take a further step to make the considered model a bit more complex. Consider the control problem for the organizational structure shown in Figure 3.15. Such structures are known as *systems with distributed control* [61]. In fact, this is a reversed fan structure, where a single agent is subordinated to several principals.

Systems with distributed control often arise in contemporary matrix and *project-oriented* management structures, where a certain agent working for a specific project is subordinated to a project manager; at the same time, the agent belongs to some department and appears subordinate to the corresponding functional manager. The same story applies to a

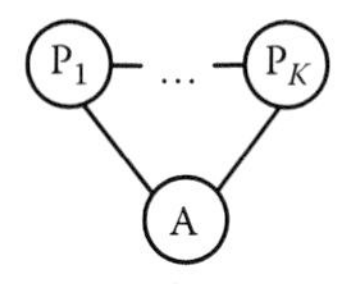

Figure 3.15 A system with distributed control.

lecturer working at a certain department of university when he is invited to deliver lectures to another department or faculty.

The incentive problem for a system with distributed control is also known as the *common agency* problem [15]. The main issue of common agency is competition among principals for the agent resulting in the game of the principals (as opposed to the fan structure that possesses a game among the agents; see Chapter 1). With several agents being added (provided that each agent is subordinated to different principals), we derive a certain game involving both sides at every level (see Figure 1.12e).

Let us describe a model that is more complicated than the previously discussed multi-agent system. Indeed, a game of agents lies in choosing certain actions (scalar variables), while that of principals consists in choosing incentive functions for the agents (depending on their actions). In other words, the game of principals includes incentive functions as their strategies. The goal functions of the principals are

$$\Phi_i(\sigma(\cdot), y) = H_i(y) - \sigma_i(y_i), \quad i \in K = \{1, 2, \ldots, k\}$$

and represent the difference between the income and the reward paid to the agent, where K designates the set of the principals.

The goal function of the agent is defined by $f_i(\sigma(\cdot), y) = \sum_{i\in K} \sigma_i(y) - c(y)$; that is, the agent receives rewards from the principals (they are summed up) and incurs some costs.

Suppose that actions of an agent belong to a multidimensional set (instead of a segment on the real axis, indicating hours or pieces). Multidimensional sets reflect actions of different types (e.g., quantity and quality of the results of agent's activity). Then the cost function maps the action set into the set of real values.

Define the agent's choice set as the set of all maxima of his or her goal function given the incentives provided by the principals:

$$P(\sigma(\cdot)) = \operatorname*{Arg\,max}_{y\in A}\left[\sum_{i\in K} \sigma_i(y) - c(y)\right].$$

Agent's behavior is rather simple—depending on the vector of incentive functions the agent chooses an action maximizing his or her goal function (the difference between the total reward and the costs).

The principals should assign payment schemes to the agent. Note each principal has to make an independent decision on how he or she would control the agent and what reward to promise to the agent. Hence, the principals are "bound" to a single agent; what a principal would actually do depends on the offer of every principal.

Every principal should not think of the strategy profile separately; in particular, if a principal asks an agent to do something, the latter may not necessarily choose the required action. The reason is that another principal may ask the agent to perform a different action and promise a higher reward to the agent. Therefore, the principals participate in a game and should arrive at an equilibrium, choosing proper incentive functions and making forecasts regarding the agent's actions given the incentive functions vector chosen.

The formulated problem is rather cumbersome; let us discuss a series of known ways to simplify it.

The first result states the following. With a game of the principals being considered, game theory suggests two common approaches to solve it, notably, the concepts of Nash equilibrium and Pareto efficiency. Systems with distributed control are remarkable for the fact that Nash equilibria set in the principals' game intersects the Pareto set; in other words, it is possible to choose a Pareto-efficient set from the set of Nash equilibria. One can prove that there exists a class of simple incentive functions that guarantee Pareto efficiency for a Nash equilibrium in the game of principals. These incentive functions have a compensatory form:

$$\sigma_i(x, y) = \begin{cases} \lambda_i, & y = x \\ 0, & y \neq x \end{cases}, \quad i \in K.$$

This incentive scheme means the principals agree to motivate the agent choosing a certain action (the plan x) and pay the agent only when he or she chooses this action. Principal i pays the amount λ_i for fulfilling the plan. Hence, this result allows for passing from the game of the principals (the corresponding strategy lies in choosing a function) to a simpler game (where a strategy consists in selecting an action of the agent and payment amounts). The following should be noted concerning the payment vector. The goal function of the agent includes the total incentive obtained by the agent and the costs incurred. If the costs vanish at zero point, then the total incentive should be not smaller than the costs $\sum_{i \in K} \lambda_i \geq c(x)$ for the agent to choose the action x.

On the other hand, only those rewards that could not be reduced without modifying the agent's action are Pareto efficient for the agent. Hence, the total incentive should be exactly equal to the agent's costs.

This fact helps to describe the equilibrium in the game of the principals. For this introduce the following notation: $W_i = \max_{y \in A}[H_i(y) - c(y)], i \in K$. If principal i would be the only principal paying the agent, he or she would involve a compensatory incentive scheme; the resulting profit of the principal would make W_i (this follows from the solution to the basic incentive problem discussed previously).

Next we formulate conditions when every principal benefits more from interaction with the other principals (*joint control* of the agent) than in the case of individual behavior (when he or she says, "The agent works only for me."). This condition is rewritten as $H_i(x) - \lambda_i \geq W_i, i \in K$. When the principals interact with each other, principal i gains the income $H_i(x)$ (as the agent chooses the action x), as well as pays the amount λ_i to the agent. At the same time, the value of his goal function should be not less than if he or she was the only principal to pay the agent (this would yield the utility

W_i). In addition, the total incentive of the agent should be equal to his costs. Denote by

$$\Lambda = \left\{ x \in A, \vec{\lambda} \geq 0 \middle| \sum_{i \in K} \lambda_i = c(x),\ H_i(x) - \lambda_i \geq W_i,\ i \in K \right\}$$

the set of agent's actions x and payment vectors $\vec{\lambda}$ for his or her activity (paid by the principals) such that the total reward exactly equals the agent's costs and each principal gains the payoff not smaller than as if he or she acted alone. This domain is a subset of the Cartesian product of the set A and the k-dimensional positive orthant.

The set Λ represents a *domain of compromise* for the system with distributed control. In a certain sense, it resembles the domain of compromise in the basic incentive problem with the single principal.

Assertion 3.3 [61, 127]

1. If the domain of compromise Λ is nonempty, *cooperation among the principals* takes place, i.e., the principals may negotiate (a) the action vector to be chosen by the agent and (b) the payment share.
2. The domain Λ may be empty. This is the situation of *competition among the principals.*

In the latter case, the result of the game involving the principals will have the following practical interpretation. The principals have failed to negotiate the action of the agent. Then principal 1 decides what he or she would like to gain from the agent (the principal acts independently). Other principals follow the example. Each principal tells the agent, "You will work for me; that is, choose a certain action and I will pay you a reward." The principal starts with costs compensation. As soon as all principals have made their proposals, the agent "is stranded." After that, a certain principal has a brain wave, suggesting, "I will cover your costs and give a bonus if you work for me only." This offer is better for the agent, since he or she obtains something in addition to the compensated costs. Then the principals start competing with each other to gain the agent over. This being the case, the agent holds the best position. The principal having the largest W_i will definitely win (this parameter describes the profit gained by the principal from the agent's actions). The agent will be "monopolized" by the principal possessing the most efficient interaction with the former.

Assume that the principals are arranged in the decreasing order of W_i, as follows: $W_1 \geq W_2 \geq \ldots \geq W_k$. Then the winner will be a principal with a maximal W_i; in addition to the compensated costs, he or she will pay W_2 plus an infinitely small value to the agent (to gain the agent over from the second principal).

The situation with efficiency-based arrangement of the principals (when the winner is the one possessing the maximal efficiency, and the winner's price exceeds the price suggested by the next principal in the ordered sequence) is referred to as an *auction solution* in a second-price auction, also known as the Vickrey auction [85, 116].

Let us formulate the existence conditions for the cooperation mode. Introduce a certain parameter connected with the maximal total payoff of the principals; in other words, define the agent's action attaining the maximum to the total profit of the principals minus the agent's costs:

$$W_0 = \max_{y \in A} \left[\sum_{i \in N} H_i(y) - c(y) \right].$$

Assertion 3.4 [61, 127]

The cooperation mode is feasible (i.e., the domain of compromise is nonempty) if and only if the total payoff of the principals (resulting from their independent activity) does not exceed the total payoff of the system under joint actions of the principals:

$$\Lambda \neq 0 \Leftrightarrow \sum_{i \in K} W_i \leq W_0.$$

The interpretation of Assertion 3.4 is as follows. The system should possess the *emergence property* (the whole is greater than the sum of its parts). In the present case, the whole (cooperation of the principals) should be greater than the sum of its parts (independent behavior of the principals). That is, if the system includes the synergetic effect, the principals will be able to reach a compromise.

In other words, Assertion 3.4 says that cooperation of the principals is possible if, and only if, common agency is economically advantageous, that is, if and only if having a single agent turning among the principals is cheaper (in terms of the total costs) than endowing every principal with his or her own "copy" of the agent (this inherits the cost function and the feasible actions domain of the latter). In practice, this is the case only when a single principal (e.g., a project manager) cannot give a full load to a worker and the costs of switching among the tasks are not appreciable.

TASKS AND EXERCISES

3.1. An organizational system consists of a single principal and a single agent. The cost function of the agent and the income function of the principal are given by $c(y) = y^2/(2r)$, $y \geq 0$ and $H(y) = \alpha \sqrt{y}$, respectively. Derive goal functions of the principal and agent.

3.2. Within the framework of Task 3.1, but assuming that $H(y) = y$, perform the following tasks:

3.2.1. Solve the incentive problem under the hypothesis of benevolence. Draw the domain of compromise. Analyze the consequences of rejecting the hypothesis of benevolence.

3.2.2. Establish the relationship between the domain of compromise and the reservation wage of the agent.

3.2.3. Solve the incentive problem if $y \in [0; A^+]$, where A^+ designates a positive constant.

3.2.4. Solve the incentive problem when the reward is bounded from above by some $R > 0$.

3.2.5. Find the optimal incentive schemes of J-type, L-type, D-type, LL-type, L+J-type.

3.2.6. Write down the minimum constraints for the incentive mechanism to implement the given action.

3.2.7. Does a solution to the optimal planning problem depend on the bonus δ?

3.2.8. Under the assumptions introduced in Section 3.1, demonstrate that maximization in the optimal plan formula (see Section 3.2) may be performed over the complete positive semi-axis (instead of the set S).

3.3. For a concave cost function of the agent, find an optimal incentive scheme of L-type.

3.4. Demonstrate the equivalence of the following goal functions: "the incentive minus the costs" and "the income minus the penalties." Use Task 3.2 as an example.

3.5*. [127] Suppose an agent has two feasible actions, $A = \{y_1; y_2\}$ and $A_0 = \{z_1; z_2\}$; the corresponding probabilities are defined by the matrix

$$P = \begin{Vmatrix} p & 1-p \\ 1-p & p \end{Vmatrix}$$

where $1/2 < p \leq 1$. The costs incurred by choosing the first and the second actions constitute c_1 and c_2, respectively ($c_2 \geq c_1$). The expected income of the principal gained by the first (second) action makes H_1 (H_2, respectively). Evaluate an optimal contract.

3.6*. Consider the dilemma of balancing between the income (q) and leisure time (t). Assume that the goal function is given by $u(q, t) = \beta\, q\, t$, $t \in [0; 16]$, and the unearned income is equal to q_0. Find optimal working time in the view of an agent. Analyze the income effect and the replacement effect.

3.7. Study comparative efficiency of the incentive schemes mentioned in Task 3.2.5.

3.8. Plot the relationship between the income and leisure time for Task 3.2.5.

3.9. Provide examples of different relationships between the desired working time and the wage rate.

3.10. Interview several persons; based on their answers, plot the relationship between the desired working time and the wage rate. Give a practical interpretation of the results in terms of the incentive problem.

3.11. [127] An organizational system consists of a single principal and several weakly related agents. The cost functions of the agents are $c_i(y_i) = y_i^2/(2r_i)$, $i \in N = \{1, 2, \ldots, n\}$, while the income function of the principal is defined by $H(y) = \sum_{i \in N} \alpha_i y_i$ (where $\{\alpha_i\}_{i \in N}$ are positive constants). Solve the incentive problem under the budget constraint imposed on the wages fund. Evaluate the optimal value of the wages fund.

3.12. [127] Solve the incentive problem for a system with a single principal and two agents with the cost functions

$$c_i(y) = \frac{(y_i + \alpha\ y_{3-i})^2}{2r_i}, \quad i = 1, 2;$$

here α is a parameter describing the level of agents' interdependence. The income function of the principal is $H(y) = y_1 + y_2$, while the wages fund is bounded above by R.

3.13. [127] There exist two agents with the cost functions $c_i(y_i) = y_i^2/(2r_i)$, $i = 1, 2$. The principal's income function represents the sum of agents' actions. Individual rewards of the agents satisfy the independent constraints $d_1 \leq \sigma_1 \leq D_1$ and $d_2 \leq \sigma_2 \leq D_2$ (i.e., there are "wage brackets"). In addition, the global constraint $\sigma_2 \geq \beta\, \sigma_1$ takes place (agent 2 possesses a higher skill level than agent 1: $r_2 \leq r_1$; hence, the former gains a greater reward than the latter for performing the same actions: $\beta \geq 1$). Find an optimal solution to the incentive problem.

3.14. Find a Nash equilibrium of the game involving a single principal and n agents ($n \geq 2$). The goal function of agent i, $f_i(y, r_i)$, is the difference between his income $h_i(y)$ (gained by the joint activity) and the costs $c_i(y, r_i)$: $f_i(y, r_i) = h_i(y) - c_i(y, r_i)$, $i \in N = \{1, 2, \ldots, n\}$. Here r_i indicates a type of agent i, representing efficiency of the agent. The income function and the cost function of agent i are defined by

$$h_i(y) = \gamma_i\, \theta\, Y, \quad i \in N, \quad \text{and}$$

$$c_i(y, r_i) = \frac{y_i^2}{2\left(r_i \pm \beta_i \sum_{j \neq i} y_j\right)}, \quad i \in N,$$

$$\left(Y = \sum_{i \in N} y_i, \sum_{i \in N} \gamma_i = 1\right),$$

respectively. It is assumed that $\sum_{j \neq i} y_j < \frac{r_i}{\beta_i}$ in the case of negative sign in the denominator. Solve the incentive problem for two agents under the assumption that the principal uses proportional incentive schemes for every agent with the individual rates λ_1 and λ_2.

3.15*. [127] Find an optimal lump-type incentive scheme of the form

$$\sigma_i(y_i, y_2) = \begin{cases} C_i, & y_1 + y_2 \geq x \\ 0, & y_1 + y_2 < x \end{cases}$$

for two agents having the cost functions $c_i(y_i) = y_i^2/(2r_i)$; here r_i designates a type of agent i, $y_i \in \Re_1^+$, $i = 1, 2$.

3.16. Find an optimal incentive scheme for motivating the agents based on the results of their collective activity: $z = \sum_{i \in N} y_i$ provided that $H(z) = z$. For agent i, the cost function is defined by $c_i(y_i) = y_i^2/(2r_i)$, $i \in N$. The principal's income function is given by $H(z) = z$.

3.17. Solve Task 3.16 provided that

$$c_i(y) = \frac{(y_i + \alpha\, y_{-i})^2}{2r_i}, \quad i = 1, 2.$$

3.18. [127] Under the conditions of Task 3.16, find an optimal *unified incentive scheme* (the same for all the agents).

3.19*. [127] Find an optimal unified rank-type incentive scheme (see also Section 4.7) for three agents; the principal's income function is the sum of agents' actions, while the cost functions of the agents are $c_i(y_i) = k_i y_i$, $k_1 > k_2 > k_3$.

3.20*. Under the conditions of Task 3.19, find an optimal competition rank-type incentive scheme (see also Section 4.7).

3.21*. [127] Find an optimal number of identical agents with the cost function $c(y) = y^2/(2\beta)$, if the principal's income is proportional to the sum of agents' actions.

How would the optimal solution change if the principal's goal function is multiplied by a decreasing function of the number of agents? Give examples of such functions and analyze them.

3.22*. Solve Task 3.21—give the examples and analyze them—under the following assumption. Under a fixed set of the agents, the principal defines the subset of agents (being included in the system staff) and guarantees a given level of utility for them. For the remaining agents (that have not been included in the staff), the principal assigns a different level of utility.

3.23. Find the domain of compromise in an organizational system with distributed control. The corresponding functions are $c(y) = y$, $H_i(y) = \alpha_i\, y$, $\alpha_i \geq 1$, $i \in K$, $k = 2$.

3.24. Find the domain of compromise in an organizational system with distributed control. The corresponding functions are $k = 2$, $c(y) = y^2$, $H_1(y) = \beta - \alpha_1\, y$, $H_2(y) = \alpha_2\, y$.

3.25. Find the domain of compromise in an organizational system with distributed control. The corresponding functions are $k = 2$, $c(y) = y$, $H_i(y) = y - y^2/(2r_i)$, $i \in K$.

3.26*. Within the framework of Task 3.24, analyze reasonability of introducing an additional level of control (a "meta-principal").

3.27*. [127] An organizational system consists of two participants with the goal functions $W(z, y) = y - z$ and $w(z, y) = z - y^2$. The participants choose actions $z \geq 0$ and $y \geq 0$, respectively. Evaluate their payoffs in the hierarchical games Γ_0–Γ_3 (see Chapter 1) provided that participant 1 moves first. What changes if participant 2 makes the first move?

3.28*. An organizational system consists of two participants with the goal functions $f_i = y_i + \alpha_i\,(1 - y_{-i})$, $y_i \in [0; 1]$, $i = 1, 2$. Evaluate their payoffs in the hierarchical games Γ_0–Γ_3 (see Chapter 1): (a) with side payments and (b) without side payments. Consider situations when the first move is made by participant 1 and by participant 2.

3.29*. [24] Prove that if a penalty function is concave on the domain of feasible agent's actions then it satisfies the "triangle inequality."

3.30*. Give the following definitions and illustrative examples for the following terms [127]:

Incentive function
Hypothesis of benevolence
Bonus
Domain of compromise
Principle of costs compensation
Decomposition principle
Aggregation principle
Incentive scheme:
- Jump-type incentive scheme
- Compensatory incentive scheme
- Linear incentive scheme
- Derived incentive scheme
- Composite incentive scheme

Optimal plan
System with weakly related agents
Budget constraint
Distributed control system
Cooperation mode
Competition mode

Chapter 4

Planning Mechanisms

Similar to what we did in Chapter 3, here we return to the problems of motivational control (where control action changes goal functions of the agents). Notably, we study a class of problems referred to as *planning mechanisms*. The term *planning* is used here in two contexts. First, a plan represents a course of actions. Second, in a narrow sense, a plan is a desired (by the principal) state of a control object. In the theory of control in organizations, a planning mechanism is a procedure of calculating the plan from agents' messages (requests, bids).

Consider the following example. The principal has to make a decision as to how to distribute $100 million between two development projects (equipment purchase and building renovation) taking into account interests of three departments (A, B, and C). Let the desired distributions for each department be described by the amounts of money allocated to equipment purchase in Table 4.1. If the principal knows for sure the distributions desired by each department, then the reasonable distribution is to take the average of these three values. Then, the decision will be to spend (20 + 40 + 60)/3 = $40 million to purchase new equipment.

But how should the principal allocate money if the distributions desired by departments are private information of each

Table 4.1 Interests of Departments

Department	*A*	*B*	*C*
Money for equipment purchase ($ millions)	20	40	60

department or information shared among departments but not available to the principal? The first possibility is to ask agents directly: "What is the best division of money between two development projects from the point of view of your department?" But the open question is whether the agents will report the requested information truthfully or not.

For instance, if agent C reports that $100 million should be allocated to equipment purchase but others tell the truth, then the final distribution will be (20 + 40 + 100)/3 = $53.3 million to purchase new equipment. Evidently, this result is more beneficial to agent C (in contrast to the result yielded by truth-telling).

This is an example of planning mechanism—the principal determines the plan as the average of agents' opinions. This mechanism is not *strategy-proof;* for some agents telling the truth is not compatible with their incentives and they are better off acting strategically.

The problem is that many mechanisms yield the best solution to some planning problems in the case when agents do not act strategically but are not strategy-proof. It was initially formulated in the 1960s by L. Hurwich [76], who introduced the notion of *incentive compatibility*. He wrote the pioneering works on *mechanism design*, which now incorporate theoretical results from a wide variety of theories:

- Auction theory by Vickrey et al. [85, 108, 118, 137, 150]
- Theory of contacts by Holmstrom et al. [65, 69, 112, 119]
- Strategy-proof social choice by Arrow et al. [4, 59, 114, 117, 119, 142, 143]; in the former USSR, by Aizerman et al. [2, 47]
- Implementation theory by Maskin et al. [44, 48, 49, 66, 77]

In the former USSR the first version of the condition to which a planning mechanism must comply to be incentive compatible was formulated in 1971 by V. Burkov and A. Lerner [34] and gave rise to the *theory of active systems* [20, 30–33].

The main results of basic planning mechanisms research are studied in this chapter.

4.1 Incomplete Information in Organizational Systems

In Chapter 1, decision-making models have been studied under the hypothesis of rational behavior (i.e., control subjects maximize their goal functions via a proper choice of available actions). In addition, we adopted the hypothesis of determinism: the subject seeks to eliminate the existing uncertainty and make decisions under complete information. For instance, when choosing a plan, a principal should make decisions (following the hypothesis of determinism) only after all uncertainty is eliminated. What does the uncertainty mean? In fact, it is insufficient awareness on essential characteristics of the environment or controlled agents. Obviously, the latter are often better informed about their characteristics than the principal is; in other words, they have some *private information*. Hence, if the principal does not possess enough information to make a decision, he or she has several alternatives to eliminate the uncertainty.

A possible way is to employ the principle of maximal guaranteed result, when the principal expects the worst-case values of the agents' parameters. However, the following idea comes to mind. If the agents have more information about something than the principal, then the principal should ask them. Hence, the principal asks, the agents report some information, and he or she makes a decision. On the other hand, the agents follow their preferences (in particular, certain control decisions of the principal could be preferable to them). Therefore, being provided with the opportunity to influence decisions of the principal (through their messages), the agents would attempt to report information to make the principal choose a decision being the most beneficial to them. In other words, information reported by the agents would not be definitely true.

The stated phenomenon of data manipulation is called the *effect of strategic behavior*. The question is immediately, "What decision-making procedures appear *strategy-proof*?" Reformulating it, "What decision-making procedures motivate the agents to reveal true information?" It would be desirable to use decision-making rules such that the agents benefit from truth-telling. This problem is considered in the present chapter.

Formally, one may draw the following analogy between incentive problems (discussed in Chapter 3) and planning problems (see Table 4.2).

Under asymmetric information, almost any control problem in organizational systems can be treated as a planning problem. Note that a solution to a corresponding control problem defines a plan being assigned to the agents for each feasible vector of their messages.

In Section 1.2 basic methods of informational uncertainty elimination were listed; these methods can be applied by a principal. Depending on information on the state of nature (being available to the principal at the moment of decision

Table 4.2 Incentive Problems versus Planning Problems

	Incentive Problems	*Planning Problems*
Strategy of an agent	$y \in A'$	$s \in \Omega$
Control	$\sigma(y)$	$\pi(s)$
Preferences of an agent	$f(y, \sigma(\cdot))$	$\varphi(s, \pi(\cdot))$

making), the following types of uncertainty are often identified:

1. *Interval uncertainty* (the principal only knows a set Ω of possible values possessed by the state of nature)
2. *Probabilistic uncertainty* (the principal knows probability distribution of the states of nature over the set Ω)
3. *Fuzzy uncertainty* (the principal knows the membership function for different values of the state of nature, defined on the set Ω)

Most of the models discussed in this chapter are the models of interval uncertainty (except the probabilistic models of contract theory).

4.2 Revelation of Information

Consider a two-level multi-agent OS composed of a single principal and n agents. A strategy of every agent lies in reporting certain information $s_i \in \Omega_i$, $i \in N = \{1, 2, \ldots, n\}$ to the principal. Based on the reported information, the principal assigns the plans $x_i = \pi_i(s) \in X_i \subseteq \Re^1$ to the agents; here π_i: $\Omega \to X_i$, $i \in N$ is a planning procedure (mechanism), $s = (s_1, s_2, \ldots, s_n) \in \Omega = \prod_{i \in N} \Omega_i$ means a vector of the agents' messages. In different settings the plans may have different nature—these may be financial or material resources allocated among agents, assigned corporate manufacturing orders, or a collective judgment (e.g., that of new plant placement or dividend appointment).

The *preference function* $\varphi_i(x_i, r_i)$: $\Re^2 \to \Re^1$ of agent i reflects his or her preferences in planning problems; this function depends on the corresponding component of the plan (assigned by the principal) as well as on a certain parameter (i.e., a *type* of the agent). As a rule, one understands the agent's type as a maximum point of his or her preference function (in other words, this is the most beneficial value of the plan in the agent's view).

When making decisions, each agent is aware of the following information: the planning procedure, the value of his or her type $r_i \in \Re^1$ (known as an *ideal point,* a *peak point,* or a *top*), the goal functions, and the feasible sets of all agents. The principal knows the functions $\varphi_i(x_i, \cdot)$ and the sets of messages available to the agents, yet the principal possesses no information on the exact values of the agents' types. The move sequence is as follows. The principal chooses a planning procedure and announces it to the agents; being aware of the planning procedure, the latter report to the former the source information for the planning procedure.

The decision made by the principal (the plans assigned to the agents) depends on the information reported by the agents; hence, the agents are able to take advantage of the situation and exert an impact on the decision (by reporting information ensuring the most beneficial plans for them). Naturally, in such conditions the information obtained by the principal may appear spurious. Hence, the problem of strategic behavior arises.

Generally, one studies planning mechanisms (in models of information revelation) under the assumption that the preference functions of the agents are *single-peaked* ones with the peak points $\{r_i\}_{i \in N}$. This implies that the preference function $\varphi_i(x_i, r_i)$ is a continuous function with the following properties. It strictly increases with respect to x_i, reaching the unique maximum point r_i, and strictly decreases then. The assumption means that the agent's preferences over the set of feasible plans are such that there exists a unique value of the plan, being optimal for the agent (a peak point); for any other plan, the preference level monotonically decreases as one moves off the peak point.

Suppose that the agents do not cooperate, playing dominant or Nash equilibrium strategies. Let s^* be the vector of equilibrium strategies.* Obviously, the equilibrium point $s^* = s^*(r)$ generally depends on the agents' types profile $r = (r_1, r_2, \ldots, r_n)$.

The *planning mechanism* $h(\cdot): \Re^n \to \Re^n$ is said to be the *direct mechanism, corresponding to* the mechanism $\pi(\cdot): \Omega \to \Re^n$ if $h(r) = \pi(s^*(r))$; it maps the vector of peak points of the agents into the vector of plans. Involving the term "direct" is justified by the fact that the agents report their peak points directly; within the framework of the original (indirect) mechanism $\pi(\cdot)$, they are able to reveal some indirect information. Imagine truth-telling is an equilibrium strategy for all agents in the corresponding direct mechanism; then the mechanism is called an *equivalent direct* (strategy-proof) *mechanism.*

Consider possible approaches to guarantee truth-telling. Probably, the suggestion to introduce a certain penalty system for data manipulation seems the most evident (under the assumption that the principal eventually (*ex post*) becomes aware of the actual values of the parameters $\{r_i\}_{i \in N}$). It has been demonstrated that truth-telling is ensured by adoption of "sufficiently great" penalties [32]. Suppose now that the principal does not expect ex post information about the parameters $\{r_i\}_{i \in N}$; an identification problem (based on information available to the principal) arises then for the unknown parameters. As a result, one also deals with the problem of constructing a penalty system for indirect indicators of data manipulation.

Another way of striving for true information reported consists in using *progressive mechanisms* [32], where the functions φ_i are monotonic with respect to the estimate s_i, $i \in N$ for any opponents' action profile. Clearly, if the *hypothesis of true estimates* ($s_i \le r_i$, $i \in N$) takes place, then the dominant strategy of every agent is to report the truth: $s_i = r_i$, $i \in N$.

* In the case of several equilibria, definite rules should be formulated for choosing a unique equilibrium from any set of equilibria.

The Perfect Concordance Condition: The Fair Play Principle

The *fair play principle* (FPP) is a fundamental theoretical result. The underlying idea is to use a planning procedure that maximizes goal function of every agent (under the assumption that agents do not manipulate information). In other words, the principal trusts the agents and believes they would not live up to his or her expectations. This explains another term widely used for the mechanism based on the FPP: the *fair play mechanism*. It was initially offered by Burkov and Lerner in 1971 [34]. Later it was translated into game-theory terms [32]. Here we give a formal definition in the final form introduced by Burkov and Enaleev in 1985 [23].

The condition

$$\phi_i\left(\pi_i\ (s),\ s_i\right) = \max_{x_i \in X_i(s_{-i})} \phi_i\left(x_i,\ s_i\right), \quad i \in N, s \in \Omega,$$

where $X_i(s_{-i})$ represents a set of plans (*decentralizing sets*) depending on the opponents' action profile $s_{-i} = (s_1, s_2, \ldots, s_{i-1}, s_{i+1}, \ldots, s_n)$ for agent i, is called the *perfect concordance condition*.

The planning procedure that maximizes the goal function $\Phi(\pi, s)$ of the principal over the set of plans satisfying the perfect concordance condition is referred to as the fair play mechanism.

The following result takes place. Revelation of information is a dominant strategy of the agents if and only if the planning mechanism is a fair play mechanism [23, 32]. This idea can also be formulated in terms of the *revelation principle*, which is due to the Nobel Prize winner R. Myerson [117, 119].

The aforementioned statement says nothing about the uniqueness of equilibrium strategy profile. Of course, with the condition of benevolence being met (if $s_i = r_i$, $i \in N$, appears a dominant strategy, then the agents reveal true information), using the fair play principle guarantees the agents' truth-telling.

Let us formulate a sufficient condition for the uniqueness of the equilibrium strategy profile $s_i = r_i$, $i \in N$ under a fair play mechanism. For agent i, denote by $E_i(s_i) = \operatorname{Arg}\max_{x_i \in X_i} \varphi_i(x_i, s_i)$ the set of his or her concerted plans. We will say that the condition of equitable preference functions holds for agent i if the following expression is valid [32]:

$$\forall\ s_i^1 \neq s_i^2 \in \Omega_i, \quad E_i(s_i^1) \cap E_i(s_i^2) = \varnothing.$$

That is, for all feasible noncoinciding estimates s_i^1 and s_i^2, the corresponding sets of concerted plans do not intersect. The condition of equitable preference functions for all agents is sufficient for the existence of a unique equilibrium strategy profile.

A necessary and sufficient condition for truth-telling being a dominant strategy under any agent types profile $r \in \Omega$ is, in fact, the existence of the sets $\{X_i(s_{-i})\}_{i \in N}$ satisfying the perfect concordance condition [23]. This assertion could be reformulated as follows. Assume that a dominant strategy equilibrium exists in the original planning mechanism; then the corresponding direct mechanism is strategy-proof.

A promising approach to obtaining sufficient conditions of strategy-proofness is known as the *geometrical method* suggested by S. Petrakov and D. Novikov [35]. The approach is based on analyzing configurations of the *dictatorship sets* (the *dictators* are the agents, whose plans deliver maxima to their goal functions). Such sets are composed of the values of the agents' types, so specific agents obtain plans that are strictly smaller than the optimal ones, equal to the optimal ones, and strictly greater than the optimal ones. The geometrical method has made it possible to derive a series of constructive conditions for individual and cooperative strategy-proofness of planning mechanisms in OS.

Optimality of Incentive-Compatible Mechanisms in a Single-Agent Organizational System

Till now, we have been mostly focused on the conditions of truth-telling. It is reasonable to put the question on how the strategy-proofness and the optimality of the mechanisms are interconnected. In other words, is it always possible to find a strategy-proof mechanism among the optimal ones? Or is it always possible to find an optimal mechanism among the strategy-proof ones? Answering these questions is critical for management, since truth-telling per se is a convenient property, but it is not as important as optimality of the mechanism. Therefore, in the sequel we discuss some results regarding optimality (in the sense of maximal efficiency) of strategy-proof mechanisms.

It has been shown that, in the case of a single agent, for any mechanism there exists a strategy-proof mechanism with the same or greater efficiency [23]. This fact could be explained in the following way: for a single agent, the decentralizing set is given by the whole set of his feasible plans (in other words, a single agent always has a dominant strategy).

Consider an organizational system with a greater number of agents ($n \geq 2$). The conclusion regarding optimality of the incentive-compatible mechanisms holds true merely in some special cases. For instance, similar results have been obtained for the resource allocation mechanisms, the mechanisms of collective expert decisions (the problems of active expertise) and the transfer price mechanisms studied below in the present chapter. They have been also obtained for other planning mechanisms.

The facts concerning the relation between the properties of optimality and strategy-proofness in planning mechanisms give causes for optimism; the matter is that these properties are not mutually exclusive. At the same time, a number of examples indicate (general) nonoptimality of the mechanisms ensuring revelation of information. Hence, the question of correlation between optimality and strategy-proofness is still open.

Studying the incentive mechanisms in OS (see Chapter 3), we used the term *incentive-compatible mechanism* when the agents are motivated to fulfill the assigned plan. Imagine an OS where the agents' strategies consist in choosing both the messages and actions (in fact, this is a "hybrid" incentive-planning problem). If the mechanisms are simultaneously incentive compatible and strategy-proof, they are said to be *correct mechanisms.* Of crucial importance is the issue when one may find an optimal mechanism in the class of correct mechanisms.

The Hypothesis of Weak Impact

Consider a situation when some planned rates λ in the system with several agents are the same for all agents; in other words, the plan has the form $\pi = (\lambda, \{x_i\}_{i \in N})$. We have to find a control λ being beneficial to all agents; if the principle of incentive-compatible planning is involved, a fundamental question concerning the existence of a solution is immediate.

Such issues are not natural for systems with large numbers of agents; indeed, the impact of an estimate (provided by a specific agent) on common control is small. If every agent does not account for the impact exerted by his estimate s_i on $\lambda(s)$, the *hypothesis of weak impact* (HWI) is valid [34]. Under the HWI, one has to coordinate the plans only with respect to individual variables. It has been shown that truth-telling is a dominant strategy if the HWI is valid and the planning procedure $x(s)$ meets the perfect concordance condition [32, 34].

Now, let us consider a series of classic planning mechanisms in multi-agent organizational systems, where the fair play principle guarantees optimality.

4.3 Mechanisms of Resource Allocation

Assume that a principal owns a certain resource that is required by agents. The principal should allocate the resource among the agents. If the principal knows efficiency of agents' utilization of the resource, the problem is to allocate the resource, for example, to maximize total efficiency of resource utilization. Imagine that the agents are active and the principal is not aware of how effectively they can use the resource. The principal asks the agents what amount of resources they need and how the agents are going to use the resources. If the available quantity of the resource is limited, then the agents (generally) would not give a fair answer about the required quantity; as a result, some agents will be short of the resource. What are the situations when the principal is able to suggest a certain procedure (i.e., a rule of resource allocation among the agents) that would be strategy-proof? We put the question in a different manner. What are the situations when the principal can suggest a certain procedure such that every agent benefits from truth-telling (regardless of the actually required quantity of the resource)?

Consider the *resource allocation mechanism* $\pi(s)$ with the following properties:

1. The planning procedure $\pi(\cdot)$ is a continuous and monotonous function with respect to agents' messages (monotonicity means that the more resource an agent requests, the more resource he or she is allocated with the messages of the other agents being fixed).
2. If an agent has been given some quantity of the resource, then he or she has the right to decrease his or her request and obtain any smaller quantity.
3. If the quantity of the resource (allocated among a group of agents) is increased, then every agent of the group is given (at least) the same quantity of the resource as before.

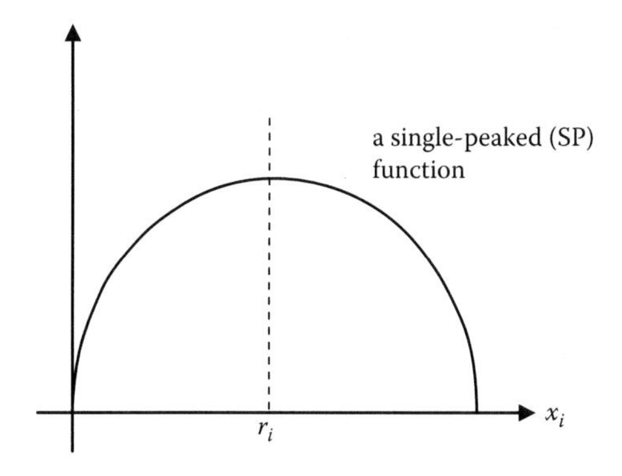

Figure 4.1 A single-peaked function.

The goal function of agent i, $f_i(x_i, r_i)$, depends on his or her type r_i which is interpreted as an optimal quantity of the resource for the given agent.

Suppose that the goal function of agent i possesses a unique maximum value over x_i in the peak point r_i. This means that some quantity is given too little (or too much), and the resulting utility of the agent decreases. Note the unique maximum value can also be reached in infinity (i.e., the goal function is a strictly increasing function). Such preference functions are known as *single-peaked* ones (see Figure 4.1).

Let us start with an example and then provide general results.

Example 4.1 [127]

Assume that $n = 3$,

$$x_i = \pi_i(s) = \frac{s_i}{s_1 + s_2 + s_3} R,$$

where R indicates the available quantity of the resource, $s_i \in [0; R]$. Set $R = 1$, $r_1 = 0.3$, $r_2 = 0.4$, and $r_3 = 0.5$. We have $r_1 + r_2 + r_3 = 1.2 > R = 1$.

1. If every agent reports the truth, the agents obtain

$$s_i = r_i \Rightarrow x_1 = 0.25; \quad x_2 = 0.333; \quad x_3 \approx 0.4.$$

2. If

$$s_i = R \Rightarrow x_i = \frac{R}{3} = 0.33 \in (r_2, r_3),$$

the first agent solves the problem:

$$\frac{s_1}{s_1 + 2} = 0.3 \Rightarrow s_1 = 6/7,$$

$$\Rightarrow s_1^* = 6/7;\ s_2^* = 1;\ s_3^* = 1 \Rightarrow x_1^* = 0.3;\ x_2^* = 0.35;\ x_3^* = 0.35.$$

This is a Nash equilibrium. •

Priority-Based Mechanisms

In *priority-based mechanisms* of resource allocation, different sorts of agents' priority indexes are used when deciding what quantity of resource each agent should get. In the general case, a priority-based mechanism is described by the following procedure:

$$x_i(s) = \begin{cases} s_i, & \text{if } \sum_{j=1}^{n} s_j \le R; \\ \min\{s_i, \gamma\,\eta_i(s_i)\}, & \text{if } \sum_{j=1}^{n} s_j > R. \end{cases}$$

Here n is the number of the agents, $\{s_i\}_{i \in N}$ designate their requests, $\{x_i\}_{i \in N}$ are the allocated quantities of the resource, R is the available quantity of the resource, $\{\eta_i(s_i)\}_{i \in N}$ is the preference functions of the agents, and γ is a certain parameter.

Minimization in the previous formula means that an agent never receives more of a resource than he or she requests.

The parameter γ serves for normalization and should fulfill the budget constraint

$$\sum_{i=1}^{n} \min\left\{s_i, \gamma\, \eta_i(s_i)\right\} = R.$$

In other words, under given requests and preference functions (in the conditions of resource deficiency), the available quantity R of the resource should be allocated in full.

Depending on the type of priority, such mechanisms could be divided into three classes: *straight priority-based mechanisms*, where $\eta_i(s_i)$ is an increasing function of the request s_i, $i \in N$; *absolute priority-based mechanisms*, where priorities of the agents are fixed and independent of the reported requests*; and *reverse priority-based mechanisms*, where $\eta_i(s_i)$ is a decreasing function of the request s_i, $i \in N$. We now discuss straight priority-based mechanisms and reverse priority-based mechanisms of resource allocation.

Straight Priority-Based Mechanisms

The procedure of resource allocation being proportional to the requests is referred to as the *mechanism of proportional allocation*:

$$x_i(s) = \frac{s_i}{\sum_{j \in N} s_j} R.$$

In fact, this is the most widespread way to distribute resources. Evidently, the stated procedure of resource

* In the case of absolute priority-based mechanisms, the plans assigned to the agents do not depend on their requests; hence, any absolute priority-based mechanism could be considered strategy-proof under the hypothesis of benevolence. This class of mechanisms has an obvious drawback—the principal uses no information reported by the agents.

allocation satisfies the normalization condition. For any combination of agents' messages, all available quantity of the resource is allocated. The conditions of continuity and monotonicity are met, as well.

Suppose that the feasible message of every agent lies within the segment [0; R]; that is, an agent may (at least) refuse from the resource and (at most) ask for the complete quantity of the available resource.

If an agent is given a certain quantity of the resource, then he or she may get a smaller quantity (down to zero) by reducing the requested quantity because of the continuity and monotonicity of the resource allocation procedure.

When every agent reports the truth (i.e., the actual quantity required), all agents are definitely given less; such a situation seems reasonable, since deficiency of the resources results in "proportional cuts" in the case of truth-telling.

Now, suppose that the game is repetitive. At step 2, the agents ask for more resource. If each agent makes the maximal request possible, all of them are given equal quantity R/n of the resource. If an agent needs less than R/n, he or she may contribute to other agents, but one or more agents will always be short of resource, as deficiency takes place.

The described mechanism is not strategy-proof because the agents benefit nothing from truth-telling of their types (the actual quantities required by them).

Thus, we have studied the example of the resource allocation mechanism and evaluated the equilibrium. To proceed, let us formulate the results about the straight priority-based mechanisms in a general form. For this, try to understand what properties are intrinsic to the equilibrium. The agents could be categorized as follows.

1. *High-priority agents* (*dictators*), gaining optimal values of the plan, viz. the values coinciding with their types (in terms of resource allocation, such agents are given exactly the required quantity of the resource).

2. *Unsatisfied agents*, lacking for the resource (in the equilibrium, these agents will be given less than they need, although they ask for the maximal quantity).

The following properties characterize the equilibrium.

Assertion 4.1 [127]

1. Suppose that in an equilibrium a certain agent is unsatisfied—he or she gets a strictly smaller quantity of the resource than he or she actually needs: $x_i^* < r_i$. Then, in the equilibrium, the agent will ask for the maximal possible quantity of the resource: $s_i^* = R$.
2. Suppose that in an equilibrium a certain agent requests a strictly smaller quantity of the resource than the maximum quantity: $s_i^* < R$. This means the agent gets the optimal quantity of the resource: $x_i^* = r_i$; in other words, he or she is a dictator.

Introduce the notion of the *anonymous mechanism* of decision making, which should be symmetric with respect to permutation of the agents (any permutation of agents' messages should be followed by a proper permutation of their plans). Anonymity appears a requirement of justice; for instance, in voting procedures it lies in the following. Two voters that exchange empty ballot papers at a polling station, in fact, exert no influence on the results of the poll.

Assertion 4.2 [127]

1. All anonymous mechanisms of resource allocation are equivalent in the sense that they lead to identical equilibrium quantities given to the agents under their identical preferences.
2. The mechanism of proportional allocation is anonymous.

Corollary [127]

As all anonymous mechanisms of resource allocation are equivalent, they are equivalent to the anonymous mechanism of proportional allocation.

Thus, any anonymous procedure satisfying the aforementioned requirements yields the same result. On the other hand, the mechanism of proportional allocation (being an anonymous one) has a simple form. It appears simple both for a study and for the agents: the allocated quantity is proportional to the requests.

Therefore, Assertion 4.2 states that one should not invent complicated mechanisms of resource allocation; within the class of anonymous mechanisms, it suffices to consider the mechanism of proportional allocation. In addition, the mechanism of proportional allocation turns out equivalent to the mechanism of serial resource allocation.

The *mechanism of serial resource allocation* is a direct mechanism; that is, each agent is asked how much resource he or she needs.

Suppose that the agents have submitted their messages $\tilde{r}_i$, $i = 1, \ldots, n$. Let us sort the agents in the ascending order of the messages (agent 1 has requested for the minimum quantity of the resource, …): $\tilde{r}_1 \le \tilde{r}_2 \le \ldots \le \tilde{r}_n$. Next, apply the following algorithm of serial resource allocation (initially, take $x_i := 0$, $i \in N$).

Step 1. If we can supply the quantity requested by agent 1 to everyone, then we give $\tilde{r}_1$ to all agents (if $n \cdot \tilde{r}_1 \le R$, then $x_i := x_i + \tilde{r}_1, \tilde{r}_i := \tilde{r}_i - r_1, i \in N$; $R = R - n \cdot r_1$). Otherwise, we allocate the resource among the agents in equal shares (if $n \cdot \tilde{r}_1 > R$, then $x_i := R / n, i \in N$) and terminate the algorithm.

Step 2. We exclude agent 1 from further consideration, decrease the requests of the remaining agents by r_1 and renumber them in the increasing order. Then return to Step 1 to distribute the rest of the resource.

Example 4.2(a) [127]

Set $R = 1$, $r_1 = 0.3$, $r_2 = 0.4$, and $r_3 = 0.5$. Sorting leads to $0.3 \le 0.4 \le 0.5$. Assume that all agents have reported the truth; hence, we can supply the minimum quantity (0.3) to everybody:

$$x_1 = 0.3;\ x_2 = 0.3;\ x_3 = 0.3.$$

Here are the requests after Step 1: $r_1 = 0$, $r_2 = 0.1$, $r_3 = 0.2$, and $R = 0.1$. Agent 1 is completely satisfied. Therefore, we forget about agent 1 and repeat the procedure for the agents still demanding for the resource. The residual quantity of the resource constitutes 0.1. This could not be supplied simultaneously to both agents following the demand of agent 1 (the former agent 2); that is, 0.1 could not be given to each of the remaining agents. Consequently, we should divide the residual quantity in equal parts (0.05) and allocate them.

As a result, agent 2 and agent 3 each gain resource quantity 0.35:

$$x_2 := x_2 + \frac{0.1}{2} = 0.35. \qquad \bullet$$

We have demonstrated operation of the mechanism of serial resource allocation. Obviously, the procedure terminates (at most) after n steps, where n is the number of the agents.

One would easily show that truth-telling is beneficial within the mechanism of serial resource allocation; that is, revelation of information is a dominant strategy for every agent. Readers are encouraged to prove this fact independently. In other words, the mechanism of serial resource allocation is a strategy-proof direct mechanism.

From the previous discussion, it is obvious that there is only one anonymous serial resource allocation mechanism: the *uniform rule* [145].

Example 4.2(a) (continued)

Using Example 4.2(a), we analyze whether an agent can improve his or her result by telling a lie.

Agent 1 obtains an optimal quantity of the resource; for him, there is no sense in distorting information. Suppose that agent 2 tries to change his or her message (overstate or understate the requested quantity). With agent 2 decreasing the request, the situation changes when the difference between the messages satisfies the following condition. The principal supplies the quantity requested by agent 2 and has enough quantity for this. The described difference is 0.05 and corresponds to dividing into equal parts. This means that agent 2 should ask for 0.35. If this is the case, he or she is given 0.35 (the same quantity as before, i.e., agent 2 benefits nothing from understating the requested quantity). If he or she asks for less than 0.35, he or she would get exactly the requested quantity. This is not beneficial to the agent since he or she would really need 0.4. Therefore, understating the request gives no benefit either.

Assume that the agent asks for more than 0.4; nothing changes in this case. The reason lies in an insufficient quantity of the resource available at Step 2, and the residuals are divided into equal parts between agent 2 and agent 3.

A similar technique is used to demonstrate that increasing (decreasing) the requested quantity does not change the situation for them. At the same time, further reduction of the request results in a worse resource allocation for them. •

Reverse Priority-Based Mechanisms

These mechanisms include $\eta_i(s_i)$ as a decreasing function of s_i, $i \in N$; they possess several advantages against straight priority-based mechanisms. Let us analyze a reverse priority-based mechanism with the preference functions

$$\eta_i = A_i/s_i,\ i \in N,$$

where $\{A_i\}_{i \in N}$ are positive constants (note that a reverse priority-based mechanism does not meet the condition of monotonicity). The value of A_i characterizes the losses in an OS caused by leaving agent i with no resource at all. Then the rate A_i/s_i determines efficiency of the resource utilization by agent i. This is why reverse priority-based mechanisms are also known as the *mechanisms of efficiency-proportional resource allocation* (EP-mechanisms).

Example 4.2(b) [27, 127]

Consider three agents ($n = 3$), with $A_1 = 16$, $A_2 = 9$, $A_3 = 4$, and $R = 18$. First, suppose that the agents strive for obtaining the maximal quantity of the resource. Find a Nash equilibrium outcome. It could be easily observed that the function

$$x_i(s) = \min\{s_i,\ \gamma(A_i/s_i)\}$$

attains the maximum value over s_i at the point satisfying $s_i = \gamma(A_i/s_i)$. Hence, $x_i^* = s_i^* = \sqrt{\gamma A_i}$.

Evaluate γ from the budget constraint

$$\sum_{i=1}^{n} x_i^* = \sqrt{\gamma}\sum_{i=1}^{n}\sqrt{A_i} = R.$$

In this case,

$$\gamma = \left(R\Big/\sum_{i=1}^{n}\sqrt{A_i}\right)^2.$$

For the present example, we have $\gamma = 4$; the corresponding equilibrium requests are computed using the condition

$$x_i^* = s_i^* = R\frac{\sqrt{A_i}}{\sum_{j=1}^{n}\sqrt{A_j}}.$$

They are $s_1^* = 8$; $s_2^* = 6$, $s_3^* = 4$.

Now, make sure this is a Nash equilibrium. Take agent 1; if he or she decreases the request ($s_1 = 7 < s_1^*$), then $s_1 + s_2^* + s_3^* < R$. Consequently, $x_1 = s_1 = 7 < x_1^*$. On the other hand, with $s_1 = 9 > s_1^*$ one obtains $\gamma \approx 4.5$; $x_1 = 8 \equiv x_1^*$.

It is easy to show that the derived strategies are secured ones for the agents; i.e., they maximize the payoffs under the worst-case strategies of the remaining agents [27].

If the preference functions of the agents attain maximum at the points $\{r_i\}_{i \in N}$ and $s_i^* > r_i$, then agent i will ask for (and be given) the quantity r_i. The matter is that reducing the request leads to a growing priority of an agent. The set of priority consumers of the resource is constructed exactly in this way [27]. •

4.4 Mechanisms of Transfer Prices

Consider an organizational system composed of a single principal and n agents. Goal function of agent i represents the difference between a linear reward (paid by the principal to agent i) and the quadratic costs that depend on the agent's action:

$$f_i(\lambda, y_i) = \lambda\, y_i - \frac{y_i^2}{2r_i}, \quad i \in N.$$

Let us study the following problem. Assume that the principal is interested in implementing an actions vector such that the sum of actions equals a given quantity R. In other words, the following constraint should be satisfied:

$$\sum_{i \in N} y_i = R.$$

For instance, the principal wants to guarantee execution of the total order R by different production units of a big corporation. The production units are supposed to manufacture the same product, and it is required to ensure the given level of the total output (note this problem has been discussed as an example in Section 2.1; here we treat it again within general analysis of planning mechanisms with strategic behavior). This is the first constraint.

In addition, we believe that the principal wants to minimize the total production costs:

$$\sum_{i \in N} \frac{y_i^2}{2r_i} \to \min.$$

However, the principal controls the agents only via a proper choice of the incentive scheme (recall a linear relationship with the slope λ between an action of an agent and his or her reward). This parameter λ, known as a *transfer price*, appears the same for all agents. With the value of λ given, the agents choose actions to maximize their goal functions. In the present case the agents are independent since their goal functions depend only on their individual actions. Therefore, the principal should choose the transfer price to minimize the agents' costs, satisfying the total action constraint and by making the agents choose their actions based on minimization of their goal functions.

To predict the choice of a rational agent, calculate the maximum point of his or her goal function. This function is concave and possesses a unique maximum. Differentiate this function to find the relationship between the action chosen by the agent and the parameter λ: $y_i^*(\lambda) = r_i\,\lambda$, $i \in N$. As the result, we derive the following problem:

$$\begin{cases} \lambda^2 \sum_{i \in N} \frac{r_i}{2} \to \min; \\ \lambda \sum_{i \in N} r_i = R. \end{cases}$$

Denote $\sum_{i \in N} r_i = H$. The posed system of equations has no free variables since the constraint uniquely determines λ; the latter would give the value of the goal function. Notably, λ

should be defined by the formula $\lambda = R/H$. Hence, the optimal value of the goal function is $R^2/(2H)$. In other words, the principal establishes full centralization in the organization; the agents are assigned plans and benefit from fulfilling them. The remaining point is to understand what plans should be assigned to the agents to attain the minimum costs of the agents (under the total output constraint). Let us obtain a solution to the stated problem.

Write down the corresponding Lagrange function (here μ indicates the Lagrange multiplier):

$$\sum_{i \in N} \frac{y_i^2}{2r_i} - \mu\left(\sum_{i \in N} y_i - R\right) \to \min.$$

Then we have $(y_i/r_i) - \mu = 0,\ i \in N,\ y_i = \mu\, r_i,\ \mu = R/H = \lambda$. Therefore, $y_i^* = r_i(R/H),\ i \in N$; in particular, the optimal action of agent i is proportional to his or her type.

Thus, we have formulated two different problems and derived identical solutions. The first problem lies in the fact that the principal should choose a transfer price such that the total costs of the agents are minimized (under the assumption that the agents choose the actions maximizing their goal functions). The second problem is to find an optimal set of the plans such that their sum equals R and the total costs of the agents are minimal. In this problem, a transfer price serves as the Lagrange multiplier ($\mu = \lambda$). Interestingly, a proportional incentive scheme has turned out optimal within the framework of the model considered. Moreover, the optimal incentive scheme has identical wage rates for all agents (such a scheme is known as *unified incentives*). Each agent could be assigned a specific wage rate; nevertheless, exactly the same wage rate for all agents appears optimal!

Let us consider the well-known problem of shortening the critical path (i.e., total execution time) of a project in progress. Assume that the critical path is invariant to a small shortening of its operations. Then the agents performing the critical operations should be paid additional rewards for reducing execution time of their operations provided that the total project's duration is reduced to a required value. Imagine that project participants performing the critical operations have the quadratic costs and are paid λ per unit reduced time. In this case, we obtain exactly the problem considered above and find its solution easily.

Of course, the obtained results (first, coinciding solutions to both problems and, second, optimality of unified incentives) take place only under the mentioned assumptions. Notably, an essential assumption of the model consists in the shape of the agent's cost function (a quadratic function of action y and a reciprocal function of agent's type (rate) r). Power-type functions yield many attractive properties in economic and mathematical models:

1. Optimality of unified linear incentive schemes (optimality of unified wage rate);
2. Feasibility of solving the aggregation problems. That is, we have studied the cost minimization problem for a given set of agents with the rates $\{r_i\}$ and obtained the following. The costs of collective execution of the order by all agents have the same form as the costs of a single agent with the type $r = H$. This means that we can replace all agents with a single agent whose action is equal to the sum of the agents' actions and whose type is equal to the sum of the agents' types. The discussed properties are intrinsic to the quadratic functions and the Cobb-Douglas functions $(1/\alpha)r_i^{1-\alpha}y_i^{\alpha}$, $\alpha \geq 1$. It could be demonstrated that the same properties are natural for functions having a general form of $r_i\,\varphi(y_i/r_i)$, where $\varphi(\cdot)$ stands for an increasing convex smooth function vanishing in the origin. The reader is asked to do it independently.

Till now, we have supposed that all parameters are known; in particular, the problem has been solved under the hypothesis that the rates r_i of the agents' cost functions (agents' types) are known to the principal. Consider the problem when the principal possesses no information on the types r_i of the agents. Denote by s_i the message of agent i about his or her type.

Based on the messages reported by the agents, the principal solves a planning problem, that is, defines the plans' vector $x(s)$; in addition, he or she determines the value of the transfer price $\lambda(s)$ depending on the agents' messages.

The obvious idea is to use solutions to the problems derived under complete information on the agents' cost functions. In other words, the principal can substitute the values of parameters reported by agent into the mechanism (the principal asks the values of just those parameters that are required to make information complete) and assign the plans according to the resulting mechanism.

Such an approach leads to the following result:

$$\lambda(s) = \frac{R}{\sum_{i \in N} s_i}.$$

The plan assigned to agent i constitutes

$$x_i(s) = \frac{s_i}{\sum_{j \in N} s_j} R, \quad i \in N$$

(we have replaced the types for the messages).

Thus, we have constructed the *mechanism of transfer prices*, which resembles the mechanism of proportional resource allocation. However, the information reported to the principal depends on the agents. Consider their goal functions and substitute the relationships $\lambda(s)$ and $x_i(s)$ there; the idea is to understand, first, whether an agent benefits from fulfilling the assigned plan and, second, what information would be beneficial for reporting. Proceeding in this way, we have

$$f_i(\lambda, s) = \frac{R^2 s_i}{\left(\sum_{j \in N} s_j\right)^2} - \frac{s_i^2 R^2}{2\left(\sum_{j \in N} s_j\right)^2 r_i} = \frac{R^2}{\left(\sum_{j \in N} s_j\right)^2}\left(s_i - \frac{s_i^2}{2r_i}\right), \quad i \in N.$$

Therefore, we have derived the goal function, which depends not on the actions, but on the messages of the agents. What messages would an agent report in order to maximize his or her goal function?

Find the maximum for the goal function of agent i over his message s_i. The denominator includes s_i, and differentiating it seems uncomfortable. This "shortcoming" is eliminated by applying the *hypothesis of weak contagion*. In particular, assume that there are sufficiently many agents; choosing a message, every agent exerts almost no impact on the transfer price (this parameter is common for all agents). In this case, the denominator of the goal function would be independent of the message of a specific agent (the total message is "constant"). Hence, we obtain $s_i = r_i$, $i \in N$; i.e., truth-telling is beneficial to all agents and the mechanism is strategy-proof. Thus, the mechanism of transfer prices possesses the following properties:

1. The agents report true information.
2. The budget constraint is fulfilled (the total action equals a given value).
3. The total costs of the agents are minimal.

4.5 Expert Mechanisms

An *expertise* is an exploration of internal properties of an object, a process, or a phenomenon by questioning experts [36, 148]. Making decisions, a principal does not possess comprehensive information on all aspects of the reality; therefore, he or she has to involve experts.

On the other hand, experts have personal preferences. It may lead to situations when an expert reports untrue information during the expertise.

This occurs in the following cases. Imagine several experts have met to make a decision in a certain field. In the course of discussion, an expert pays heed to the fact that the decision (which the experts are going to make) differs dramatically from the best one (according to his or her viewpoint). For instance, the issue of spending money is considered in a university. One of the deans believes a new supercomputer should be purchased, yet he expects that the decision regarding redecoration of the university rooms would be made. Assume this dean has previously thought that 30% of the available funds can be assigned for redecoration activities (and 70% for the supercomputers). Under modified conditions, he would definitely say the following: "I suggest spending the complete fund to purchase a new supercomputer." In other words, the dean would distort the information (by not reporting his true opinion).

The aforesaid is of crucial importance when the experts decide how finances should be distributed among them or among subjects being lobbied for (alternatively, when they prepare information for such decisions). Manipulation happens as the result of noble or mercenary motives. In the sense of mathematical modeling, it seems relevant that data manipulation may occur if each expert is interested in the expertise result (i.e., a collective decision) being as close to his or her own opinion as possible.

Suppose the result of an expertise is the quantity $x \in [d;D]$; denote by s_i the message of expert i ($s_i \in [d; D]$) and by r_i his or her true opinion ($r_i \in [d; D]$). The result of the expertise is a known function of the experts' opinions; in fact, it represents the mapping (the *expertise procedure*) $\pi(\cdot) : [d; D]^n \rightarrow [d; D]$ of the set of feasible messages into the set of feasible solutions.

Here are the conditions imposed on the expert mechanism:

1. Continuity
2. Monotonicity
3. Unanimity, that is, $\forall a \in [d;D]\ \pi(a, a, ..., a) = a$. If all experts have reported an identical opinion, it should be chosen as a collective decision.

First, let us consider an example and then state the general results.

Example 4.3

Assume that the result of the expertise belongs to the segment [0; 1] and there are three experts. Their true opinions are that the estimated quantity constitutes 0.3, 0.5, and 0.7, respectively. The expertise procedure constitutes in averaging the opinions. Such function (the arithmetical mean) satisfies all conditions listed; obviously, the arithmetic mean is continuous, monotonous, and meets the unanimity condition. Thus, we have

$$x \in [0; 1],\ n = 3,\ r_1 = 0.3,\ r_2 = 0.5,\ r_3 = 0.7,$$

$$x = \pi(s) = \frac{1}{3}\sum_{i=1}^{3} s_i.$$

The experts would act as follows. Suppose all of them have reported the truth, that is, $s_i = r_i$. Hence, the decision made is 0.5: $x(\vec{r}) = 0.5$ (this is the arithmetic mean). Let us study individual behavior of the experts. Every expert seeks to make the result of the expertise as close to his or her opinion as possible. Expert 2 is completely satisfied, as far as the result coincides with her true opinion. In contrast, expert 1 is displeased, since he needs a smaller result. Finally, expert 3 turns out discontented (he wants a greater result).

Consequently, due to the monotonous property of the function, expert 1 would decrease the message, while expert 3 would increase it. Suppose the experts report

0, 0.5, and 1, respectively. The result remains the same (0.5), so long as variations of the messages provided by experts 1 and 3 are identical:

$$s_1 = 0,\ s_2 = 0.5,\ s_3 = 1.$$

The derived message vector is a Nash equilibrium for the game of the experts. Indeed, expert 2 would not modify his message, while expert 1 would like to do it (but this is impossible – she reports the minimum value); in the end, expert 3 would like to increase the result, yet is unable (since he reports the maximal quantity). We have the same situation for other equilibria—who wants less is unable to achieve less (being "hampered" by the lower bound); on the other hand, the upper bound prevents those who want more from achieving more. •

The example demonstrates that, generally, experts (agents) reveal untrue information. The following question is immediate. What could be done to motivate them for truth-telling?

Assertion 4.3 [127] (similar to Assertion 4.1 formulated for resource allocation mechanisms)

1. Suppose that the equilibrium decision x^* is greater than the opinion of some experts, that is, $x^* > r_i$; then in the equilibrium these experts would report the minimal estimate: $s_i^* = d$.
2. Suppose that the equilibrium decision is smaller than the opinion of some experts, that is, $x^* < r_i$; then in the equilibrium these experts would report the maximal estimate: $s_i^* = D$.
3. Suppose that in the equilibrium some experts report opinions within the corresponding segment, that is, $s_i^* \in (d; D)$; then the resulting decision satisfies the experts: $x^* = r_i$.

Using Assertion 4.3 as a basis, we construct an equilibrium in the expert mechanism and analyze it. For the sake of the analysis, let us rearrange the experts in the ascending order of their true opinions:

$$r_1 \leq r_2 \leq \ldots \leq r_n.$$

Suppose that a decision has been chosen from the segment [d; D]. According to Assertion 4.3, the experts having their opinions to the left (right) of the decision would report the lower (upper) bound. Hence, the vector of equilibrium messages takes the form

$$s^* = (d, d, \ldots, d, s_k^*, D, D, \ldots, D).$$

The experts having "small" indices strive for shifting the equilibrium to the left and report minimal messages; probably, expert k – a "dictator" – reports s_k^* from the segment [d; D]. At the same time, the experts with large indices want to shift the equilibrium to the right; thus, they report maximal messages.

The equilibrium message s_k^* should be such that

$$\pi(d, \ldots, d, s_k^*, D, \ldots, D) = r_k.$$

This equation allows for finding the vector of equilibrium messages of the agents. However, the position of s_k is unknown. Indeed, how many agents report the maximal (or minimal) value and who reports the non-boundary estimate is still to be determined. If the principal knows the answers, he or she may evaluate the vector of equilibrium messages (by substituting s_k into the given equation and solving the latter).

In the previous example $k = 2$, i.e., expert 2 is a dictator. He or she wonders what his or her message should be (provided that expert 1 reports zero and expert 3 reports 1 to obtain the resulting decision of 0.5. The corresponding message should be equal to 0.5. To find index of the dictator in general case, let us introduce the following sequence of numbers:

$$w_i = \pi(\underbrace{d,\ldots,d}_{i}, \underbrace{D,\ldots,D}_{n-i}), \quad i = \overline{0, n}.$$

Fix the number of experts reporting the minimal opinions; the rest reveal the maximal ones. Varying the number of experts who report the minimal opinions from 0 to n, we derive a decreasing sequence of points. The point w_0 coincides with the right bound D (from the unanimity condition). Similarly, if all agents report the lower estimate d, the resulting decision constitutes $w_n = d$.

Thus, we have two sequences of numbers: the first one is an increasing sequence $\{r_i\}$ composed of true opinions of the experts, while the second one is a decreasing sequence of points $\{w_i\}$. We claim these sequences will definitely intersect. Let us evaluate the rightmost point of their intersection. This means one should first take the minimum of two elements having the identical index in each sequence; after that, one should find the maximum over all indices. Hence, an expert has the index

$$k = \max_{i=\overline{1,n}} \min\,(r_i, w_{i-1}).$$

In the previous example, the minimum of the opinion of agent 1 and w_1 is equal to r_1; for agents 2 and 3, these minima constitute r_2 and 1/3, respectively (i.e., the sequences $\{r_i\}$ and $\{w_i\}$ interchange for agent 3). The maximal value among these three points is 0.5. Hence, the formula gives the index of the expert, which is, in fact, a dictator (in the present example, $k = 2$).

Now, assume that a *direct expert mechanism* is utilized instead of the original π-mechanism. The direct mechanism lies in the following. The final opinion is defined on the basis of the agents' messages $\{\tilde{r}_i\}$ according to the procedure

$$\hat{x}^* = \max_i \min(\tilde{r}_i, w_{i-1}).$$

Note that the messages $\tilde{r}_i$ are a priori sorted in the ascending order.

Assertion 4.4 [127]

Under the direct expert mechanism described, truth-telling is a dominant strategy of the experts.

4.6 Basic Model of Adverse Selection

Consider an organizational system with a single principal and a single agent. The former sells a certain product to the latter; the corresponding quantity and amount of money paid are q and t, respectively. The goal function of the principal is determined by $\phi_0(t,q) = t - C(q)$.

Suppose that $C(q)$ (representing the manufacturing costs of the principal) is a twice differentiable convex function, and $C'(0) = 0$, $C'(\infty) = \infty$.

The goal function of the agent is given by $\phi_1(t,q,\theta) = u(\theta,q) - t$; here $\theta \in \Theta = [\underline{\theta}; \overline{\theta}]$ indicates a positive parameter (interpreted as a *type of the agent*). The function $u(\theta,q)$ describes utility of the product from the viewpoint of the agent; it is increasing and convex with respect to q as well as increasing with respect to θ.

The principal knows the set Θ and the probabilistic distribution of the agent's type over this set. Note that the cumulative distribution function $F(\theta)$ is differentiable, i.e., $f(\theta) = F_\theta(\theta)$.

The problem of the principal (known as *the basic problem of adverse selection*) is to maximize his or her expected utility:

$$\int_{\underline{\theta}}^{\overline{\theta}} [t(\theta) - C(q(\theta))] f(\theta)\, d\theta \to \max_{t(\cdot), q(\cdot)} .$$

Using the *revelation principle** [118], one constructs a strategy-proof mechanism, *viz.*, the *menu of contracts* $\{q(\cdot), t(\cdot)\}$ being dependent on the type estimate reported by the agent.

The necessary conditions of strategy-proofness (or *incentive compatibility*, IC-conditions) of the mechanism have the following form:

$$\begin{matrix} (IC_1) \\ \\ (IC_2) \end{matrix} \qquad \forall \theta \in \Theta, \begin{cases} \dfrac{dt}{d\theta}(\theta) = \dfrac{\partial u}{\partial \theta}(q(\theta), \theta) \dfrac{dq}{d\theta}(\theta), \\ \dfrac{\partial^2 u}{\partial q\, \partial \theta}(q(\theta), \theta) \dfrac{dq}{d\theta}(\theta) \geq 0. \end{cases}$$

Under Spence-Mirrlees's [118] single crossing property

$$\forall q, \forall \theta \quad \frac{\partial^2 u}{\partial q\, \partial \theta}(q, \theta) > 0,$$

it has been shown that the function $q(\theta)$ is nondecreasing.

Suppose that $\forall q, \forall \theta$: $(\partial u / \partial \theta)(q, \theta) > 0$. In the case of optimal strategy-proof mechanism, introduce the agent's profit function, which depends on his type: $\nu(\theta) = u(q(\theta), \theta) - t(\theta)$. We emphasize that under the condition IC_1, $\forall \theta \in \Theta$:

$$\frac{d\nu}{d\theta}(\theta) = \frac{\partial u}{\partial \theta}(q(\theta), \theta) > 0.$$

* The revelation principle is widely used and, in fact, turns out close to the fair play principle (discussed already). Both principles are equivalent for systems involving a single agent.

Therefore, the *conditions of individual rationality* of the agent ($\forall\theta\in\Theta: \nu(\theta)\geq 0$) could be ensured by $\nu(\underline{\theta}) = 0$. This implies that

$$\nu(\theta) = \int_{\underline{\theta}}^{\theta} \frac{\partial u}{\partial \theta}(q(\tau),\tau)d\tau,\ \ t(\theta) = u(q(\theta),\theta) - \int_{\underline{\theta}}^{\theta} \frac{\partial u}{\partial \theta}(q(\tau),\tau)d\tau.$$

The problem of the principal (construct a certain mechanism maximizing his or her profit) is reduced to solving

$$\frac{\partial H}{\partial q}(q^*(\theta),\theta) = 0$$

provided that

$$\frac{dq^*}{d\theta}(\theta) \geq 0. \tag{4.1}$$

Here $H(q(\theta),\theta) = \phi_0(q(\theta),t(\theta))$.

Two cases are then possible; in particular, constraint (4.1) either holds everywhere as a strict inequality or becomes the equality for specific values of θ.

The first case is simple to analyze; under condition (4.1), the corresponding Lagrange multiplier makes zero and $q(\theta)$ is evaluated from the formula $\partial H(q(\theta),\theta)/\partial\theta = 0$. In other words,

$$C_q(q) = u_q(\theta, q(\theta)) - \frac{1}{h(\theta)} u_{q\theta}(\theta, q(\theta)), \tag{4.2}$$

with $h(\theta) = f(\theta)/(1 - F(\theta))$.

An agent of each possible type is given a specific contract (the function $q(\theta)$ strictly increases with respect to θ). Hence, all types (except the largest one) obtain a certain level q that is smaller than the optimal one; the largest type obtains the efficient quantity.

Imagine a situation when constraint (4.1) holds as the equality (at least, for several values of θ); such a situation appears by far more complicated. To solve the problem, one should apply Pontryagin's maximum principle. Consider a "dynamic" maximization problem

$$\max_{q(\theta)} = \int_{\underline{\theta}}^{\overline{\theta}} H(q(\theta),\theta) f(\theta)\, d\theta,$$

where θ is a certain analog of time, q represents a phase variable evolving according to the law $dq/d\theta = \Omega$, and Ω means control being bounded from below: $\Omega \geq 0$.

As a rule, the problem is treated in the following way. First, assume that the contract is type-separating and find $q(\theta)$ from equation (4.2). If the resulting function $q(\theta)$ appears nondecreasing, then the problem is solved and the optimal *contract is type separating.* Otherwise, with the derived function possessing some segments of decrease, one has to solve the previously described optimal control problem. Note that some agents with different types are given identical contracts; that is, the equilibrium is partially *pooling.*

Example 4.4

Let $C(q) = q^2/2$ and $u(\theta,q) = \theta q$, while the agent's type is uniformly distributed over the set of feasible values:

$$f(\theta) = \frac{1}{\overline{\theta} - \underline{\theta}}.$$

Then we obtain that

$$t(\theta) = \theta q(\theta) - \int_{\underline{\theta}}^{\theta} q(\tau) d\tau,$$

and $q(\theta)$ is determined by solving equation (4.2):

$$q(\theta) = 2\theta - \bar{\theta}.$$

Having in mind the natural requirement $q(\theta) \geq 0$ (the buyer may purchase only nonnegative quantity of the product), we derive the following menu of contracts. For $\theta \in [\underline{\theta}, \bar{\theta}/2]$, this is {0,0}; for $\theta \in [\bar{\theta}/2, \bar{\theta}]$, the contract is specified by $\{q(\cdot), t(\cdot)\}$:

$$q(\theta) = 2\theta - \bar{\theta},\ t(\theta) = \theta^2 - (\bar{\theta} - \underset{\sim}{\theta})\underset{\sim}{\theta},$$

where $\underset{\sim}{\theta} = \max[\underline{\theta}, \bar{\theta}/2]$. •

Thus, it is possible to use the standard model of adverse selection to generate a flexible price schedule for a product suggested by a monopolist [141]. Offering different modifications of the product for sale at different prices (in fact, this is the contracts' menu discussed already), the manufacturer covers various groups of users. Wine provides a classical example of such product. The longer wine is allowed to age, the higher its quality is. Wine connoisseurs are willing to purchase high-quality wine for a higher price, while unsophisticated customers are satisfied with less-aged wine for a reasonable price. Applying the model of adverse selection, the manufacturer optimizes the expected income from product sales.

However, it should be emphasized that not only the quality but also the output of the product may be varied. The larger the quantity purchased by a customer, the smaller the incremental cost of the product. The relationship between the product's price and the purchased quantity is, in fact, a contracts menu that could be obtained through a standard model of adverse selection. The major issue arising in practical problems consists in identification, first, of the buyer's type and, second, of possible bounds of the type values.

4.7 Rank-Order Tournaments

Divisible-Good Tournaments

Discussing the reverse priority-based mechanisms in Section 4.3, we have underlined that the resource is allocated proportionally to the efficiency $\xi_i = \varphi_i(x_i, r_i)/x_i$ of its utilization by agents. *Rank-order tournaments* are remarkable because the resource is provided exclusively to tournament winners (the total quantity of the resource may be not enough for all agents).

Suppose the agents report two parameters to the principal: the resource request s_i; and the estimate ξ_i of the expected utilization efficiency. In this case, the total expected effect (provided to the OS by activity of agent i) constitutes $w_i = \xi_i s_i$, $i \in N$. Let us sort the agents in the descending order of their efficiency: $\xi_1 \geq \xi_2 \geq \ldots \geq \xi_n$.

Apparently, the agents may promise a lot to receive the resource. Therefore, when applying rank-order tournaments, the principal should take care of a control system to monitor the fulfillment of the commitments by the agents. Introduce a system of penalties $\chi_i = \alpha\,(\xi_i s_i - \phi_i(s_i))$, $\alpha > 0$, $i \in N$, being proportional to the deviation of the total expected effect $\xi_i s_i = w_i$ from the actual one, $\varphi_i(s_i)$. Note that the quantity $(\xi_i s_i - \varphi_i(s_i))$ characterizes deliberate deception of the agent, used to win the tournament.

The goal function of agent i has the form

$$f_i(\varphi_i, \xi_i) = \mu\varphi_i(s_i) - \alpha\left[\xi_i s_i - \varphi_i(s_i)\right], \quad i \in N,$$

where μ is a share of the effect remaining at disposal of the agents (i.e., $\mu\,\varphi_i(s_i)$ defines the income of agent i). We emphasize that the agent is penalized only in the case when $\xi_i s_i > \varphi_i(s_i)$. If the actual total effect exceeds the expected one, the penalties are set to zero.

The principal owns the resource quantity R and allocates it in the following way. Having the maximal efficiency, agent 1 is given the required quantity s_1 of the resource. The agent with the second largest efficiency obtains the resource in the second place (the corresponding quantity makes s_2); the procedure continues till the resource is completely consumed. In other words, the principal distributes the resource in the required quantity following the descending order of the efficiency rates (as long as the resource is available). The agents that have received the required quantity of the resource are known as *tournament winners*. It seems essential that some agents (e.g., the last one in the ordered sequence of winners) may have obtained less resource than they actually require. Nevertheless, these agents are still able to yield a certain effect to the principal. For the stated reasons, such tournaments are referred to as *divisible-good auctions*.

Note that the described procedure implies that winning the tournament is subject to the value of efficiency ξ_i (there is no correlation to the value of the request s_i). Thus, the agents would seek to maximize their goal functions, that is, request for the quantities attaining the maximal values to their goal function (if they win the tournament).

Denote by m the maximal index of the agent that has won the tournament (in other words, the agents with indices $j = \overline{1,\ m}$ are the winners). It could be easily demonstrated that all winners would report identical efficiency estimates in the equilibrium, that is, $\xi_j^* = \xi^*$, $j = \overline{1,\ m+1}$. Moreover, under rather general assumptions imposed on the penalty functions, rank-order tournaments ensure optimal resource allocation [21].

Discrete Tournaments

Today, one may observe the prevalence (to put it precisely, a *fashion*) of applying tournaments and auctions of every sort and kind. Various arguments are provided to justify their appropriateness. These aspects lead to the following idea. A fair competition is a real panacea for many difficult situations (or even a universal panacea), isn't it? Indeed, formal analysis of rank-order tournaments (known as *tenders* or *discrete competitions* in the case of indivisible goods) indicates that the matter is not as simple as it could seem at the first glance.

A proper definition of a tender (a discrete tournament) is given as follows. A tender represents a tournament with the property that the winners are given the exact quantity being claimed (e.g., a certain resource, finances, or beneficial project); as a result, the losers receive nothing. Efficiency of a participant is defined as the ratio of the estimated socio-economic effect (e.g., resulting from an objective expertise) to the estimate reported by the participant (e.g., a required quantity of the resource or costs). The major concept of *simple tournaments* consists, first, of sorting the participants in the descending order of their efficiency rates and, second, of serially allocating the required quantities of the resource (until the latter is completely consumed). The winners are the participants that have been given the resource. Unfortunately, the guaranteed efficiency of simple tournaments is zero [127]. (To be more precise, it could be made arbitrarily small.)

The case is somewhat better for *direct tournaments*; using the reported estimates, a tournament organizer solves the *knapsack problem* (i.e., finds a combination of the winners that is optimal in the sense of the total effect). Guaranteed efficiency of a direct tournament constitutes 0.5 [127].

TASKS AND EXERCISES

4.1. Imagine two agents (e.g., regions of a certain country, being separated by a river) provide financial resources to construct a bridge connecting both riversides. The corresponding construction costs are $c = 1$; the following *mechanism of costs allocation* is used. Each agent reports the estimate s_i of his or her "income" h_i

resulting from bridge operation. Construction of the bridge is authorized only if $s_1 + s_2 \geq c$.

4.1.1.

1. Suppose that the actual incomes of the agents are 1.4 and 0.6, respectively; in addition, let the mechanism of proportional costs allocation be used:

$$x_i(s) = \frac{s_i}{s_1 + s_2} c.$$

   Demonstrate that truth-telling regarding the incomes is not a Nash equilibrium.
2. Find all Nash equilibria.
3. Find optimal strategies provided that the agents are aware of the actual incomes of each other and a certain agent has the right to move first.

4.1.2. Under conditions of Task 4.1, suggest and analyze the mechanism of costs allocation that differs from the proportional one

4.1.3. Does an equivalent incentive-compatible mechanism (ICM) exist for the proportional mechanism of costs allocation?

4.2*. Give an example of a multi-agent organizational system with revelation of information that admits no equivalent direct mechanism [28].

4.3. Consider the incentive problem for a single-agent organizational system under incomplete awareness of the principal regarding the agent's type $r \geq 0$. The goal functions of the principal and the agent are $\varphi(y) = y - \sigma(y)$ and $f(y, r) = \sigma(y) - y^2 / (2r)$, respectively. For an arbitrary planning mechanism, demonstrate the feasibility of constructing an incentive-compatible mechanism with the same or greater efficiency [83].

What would be the efficiency of the organizational system if the principal knows true types of all agents?

4.4. The goal function of agent i is defined by

$$f_i(\lambda, x_i, r_i) = \phi_i(x_i, r_i) - \lambda\, x_i, \quad i = \overline{1,n},$$

where $\phi_i(x_i, r_i)$ represents the effect function of agent i (a convex function of the obtained resource quantity x_i), and λ indicates the resource price.

Under the hypothesis of weak contagion, demonstrate that the ICM is optimal in the sense of the total effect criterion.

4.5*. Consider a multi-agent organizational system with quasi-single-peaked preference functions of the agents. Assume that the plans assigned to the agents are monotonous functions of their messages and depend on a scalar parameter (which is chosen by the principal). Prove that for any planning mechanism of this type there exists a strategy-proof mechanism with the same or a greater efficiency [35].

4.6. An organizational system consists of the principal and five agents. The set of feasible types of the agents (i.e., the resource quantity ensuring the maximum value of the agent's goal function) is $\Omega = [0; 10]$. The principal possesses the resource in the quantity of $R = 10$. Find a Nash equilibrium for the straight priority-based mechanism

$$x_i = \begin{cases} s_i, \sum_i s_i \le R, \\ \dfrac{s_i}{\sum_i s_i} R, \sum_i s_i > R, \end{cases} \tag{4.3}$$

where s_i is the message of agent i reported to the principal. Solve the problem under the following agents' types profiles:

1. $r = \{1, 3, 5, 7, 9\}$;
2. $r = \{1, 1, 2, 8, 8\}$;
3. $r = \{5, 6, 7, 8, 9\}$;
4. $r = \{7, 8, 9, 9, 9\}$;
5. $r = \{1, 1, 2, 3, 4\}$.

4.7*. Analyze what are the fundamental differences of the equilibria described in items 1–4 of Task 4.6.

4.8. Evaluate a Nash equilibrium for the straight priority-based mechanism of resource allocation

$$x_i = \begin{cases} s_i, & \sum_i s_i \le R; \\ \min(s_i, \gamma\eta_i(s_i)), & \sum_i s_i > R, \end{cases} \tag{4.4}$$

with $\eta_i(s_i) = A_i s_i,\ \gamma : \sum_i \min(s_i, \gamma\eta_i(s_i)) = R$.

The goal functions of the agents are defined by $\phi_i(x_i, r_i) = 2\sqrt{r_i x_i} - x_i,\ i = \overline{1, n}$.

4.9*. Using the values given in items 1–4 (see Task 4.6), analyze the fundamental differences between the equilibria that correspond to the mechanisms of resource allocation given by (4.3) and by (4.4).

4.10. Study the efficiency of the following resource allocation mechanism: $s_i = \min(s_i, A_i\gamma(s_i))$, where $\gamma(s) : \sum_i x_i = R$ (provided that $\sum_i r_i > R,\ A_i > 0$). Check strategy-proofness of the stated mechanism.

4.11. Demonstrate the equivalence of all anonymous mechanisms of resource allocation that satisfy the assumptions introduced in Section 4.3.

4.12. Given the mechanism of active expertise

$$\pi(s) = \frac{1}{n}\sum_{i=1}^{n} s_i$$

with five experts, evaluate a Nash-equilibrium outcome. Suppose that the set of feasible requests of the experts is $\Omega = [10,20]$ and true opinions of the experts are the following:

1. $r = \{10, 10, 15, 20, 20\}$;
2. $r = \{10, 12, 13, 17, 18\}$;
3. $r = \{15, 15, 16, 19, 20\}$.

Prove Assertion 4.4 and use this task as a corresponding example.

4.13*. Consider the mechanism of active expertise $\pi(s)$, being optimal in the sense of closeness to the arithmetic mean

$$\pi^0(r) = \frac{1}{n}\sum_{i=1}^{n} r_i.$$

Demonstrate that it consists in partitioning the segment $[d; D]$ into n equal parts.

The goal functions of the experts are defined by $\phi_i(x, r_i) = -|x - r_i|$, $i = \overline{1,n}$. The true opinions of the experts are $r_i \in [d; D]$, $i = \overline{1,n}$. The estimates reported by the experts are given by $s_i \in [d; D], i = \overline{1,n}$.

A mechanism of active expertise $\pi^*(s)$ is optimal in the sense of closeness to the expertise $\pi^0(r)$ if

$$\max_{r \in [d,D]} |\pi^*(s^*) - \pi^0(r)| \to \min.$$

Here s^* denotes the equilibrium message vector of the experts.

4.14*. Construct the sequence $\{w_i\}$ and derive the formula of an equivalent direct mechanism for the active expertise that is optimal in the previously discussed sense (see Task 4.14) to the following expertise:

$$\pi^0(s) = \sum_{i=1}^{n} \alpha_i s_i, \text{ with } 0 \le \alpha_i \le 1, \sum_{i=1}^{n} \alpha_i = 1, \; s_i \in [0, 1].$$

4.15*. Given the mechanism of active expertise with two experts:

$$x = \sqrt{\frac{1}{n}\sum_{i=1}^{n} s_i^2}, \; s_i \in [0, 1], \; r_i \in [0, 1], \; i = \overline{1, 2},$$

construct dictatorship sets on the vector plane $r = (r_1; r_2)$ of peak points for the agents' goal functions.

4.16. Consider an organizational system composed of n agents with the Cobb-Douglas cost functions (the corresponding parameters are $\alpha = 2$, $r = 1$). Suppose that the principal pays a reward to an agent proportionally to the resulting volume of work: $\sigma_i = \lambda\, y_i$. The total volume R_0 to be fulfilled is a fixed quantity.

Construct the allocation mechanism for the volume of works based on transfer prices. For any agent, evaluate the prices for the volume of works depending on the agent's request.

Check strategy-proofness of the mechanism of transfer prices in the following cases: (a) the hypothesis of weak contagion is rejected; and (b) the hypothesis of weak impact holds true.

Assume that the cost functions of the agents appear linear or convex. How would the results change?

4.17. There is an organizational system consisting of three agents and a principal. The corresponding cost functions of the agents are

$$c_i(y_i, r_i) = \frac{y_i^2}{2r_i}\ r_i \in \Omega = [0, 1],\ i = \overline{1, 3}.$$

The principal needs the agents to fulfill the amount of work $R = 1$. Do the following:

1. Construct the mechanism of transfer prices;
2. Find Nash-equilibrium requests of the agents;
3. Estimate the efficiency of the transfer prices mechanism.

The agents' type vector is $r = \{0.3, 0.6, 0.8\}$, and the principal merely knows the set of feasible types of the agents, Ω.

Assume that the principal knows true types of the agents. What would be the system efficiency?

4.18*. Represent the resource allocation problem as an exchange problem; construct the corresponding exchange mechanism [83].

4.19*. Represent the incentive problem as an exchange problem; construct the corresponding exchange mechanism [83].

4.20*. Consider the exchange problem in an organizational system including a single agent and a single principal. Construct an incentive-compatible mechanism $\pi(s) = (x_1(s), x_2(s))$, assuming that the principal has incomplete awareness of the agent's type $r \in [r_{min}, r_{max}]$, $r_{min} > 0$. Take $f_0(x_1, x_2) = x_2 - x_1$ and $f_1(x_1, x_2, r) = x_1 - (x_2^2/2r)$ as the goal functions of the principal and agent, respectively.

The principal's problem is to maximize the expected utility of the exchange: $Ef_0(\pi(s)) \to \max\limits_{\pi(s)}$.

The principal knows that the agent's types possess a uniform probability distribution on the segment

$$[r_{\min}, r_{\max}] : F(r) = \frac{r - r_{\min}}{r_{\max} - r_{\min}}.$$

The whole resource of the first (second) type is kept by the principal (by the agent, respectively); both resources are unlimited [83].

4.21. A principal organizes planning of the volume of works in a department according to the following planning mechanism: $s_i \in [0; 1]$, $x_i > 0$, $g_1(s) = s_1 + \alpha(s_2 + s_3)$, $g_i(s) = \beta s_1 + s_i$, $\alpha > 0$, $\beta < 1/4$. Note that agent 1 is the head of the department, while agents $i = 2, 3$ are the subordinates. The parameters α and β describe the impact of increased requests on the planned work of the principal and vice versa.

Establish conditions that ensure the existence of an equivalent direct mechanism for the planning mechanism considered.

4.24*. Consider an organizational system with two agents. Planning mechanism has the following form:

$$g_1(s) = s_1 + \cos\left(\frac{3\pi}{2} s_2\right),\ g_2(s) = s_1 + s_2,\ s_i \in [0, 1],\ i = \overline{1, 2}.$$

Show that constructing an equivalent direct mechanism is impossible.

Find the set of feasible messages of the agents (being as close to the original set as possible) that allows for constructing an equivalent direct planning mechanism [35].

4.25. Demonstrate that the mechanism of serial resource allocation is an equivalent direct mechanism for the anonymous mechanism of proportional resource allocation.

4.26[*]**.** Check strategy-proofness and evaluate efficiency of the reverse priority-based mechanisms.

4.27[*]**.** Give the definitions and illustrative examples for the following terms:

Planning mechanism
Strategic behavior
Strategy-proof mechanism
Direct mechanism
Preference function
Type of an agent
Fair play principle
Decentralizing sets
Perfect concordance condition
Hypothesis of weak contagion
Single-peaked function
Mechanism of proportional allocation
Dictator
Anonymous mechanism
Mechanism of serial allocation
Mechanism of transfer prices
Transfer price
Cobb-Douglas function
Expert mechanism
Spence-Mirrlees single-crossing condition
Incentive-compatibility condition
Condition of individual rationality
Rank-order tournament
Auction:
  Continuous auction
  Discrete auction
  Direct auction
  Simple auction

Chapter 5

Informational Control Mechanisms

According to the definition given in Chapter 1, *control* is an impact exerted on a controlled system to ensure the desired behavior of the latter. The staff and structure being fixed, the controlled system represents a set of rational agents making independent decisions on how to act depending on the situation. Within the game-theoretic framework a controlled system is defined by the set N of the agents, the tuples of their goal functions $(f_i(\cdot))_{i \in N}$, and their feasible sets $(X_i)_{i \in N}$, as well as by the *awareness structure* I (also known as an *informational structure* or a *hierarchy of beliefs*). Hence, controlling a fixed set of agents may include exerting an impact on the following components: goal functions (*motivational control*), feasible sets (*institutional control*), and the awareness structure (*informational control*). This chapter discusses informational control in detail. Let us start with the following example.

Example 5.1

Consider three agents with goal functions of the form

$$f_i(\theta, x_1, x_2, x_3) = (\theta - x_1 - x_2 - x_3)x_i - \frac{x_i^2}{2},$$

where $x_i \geq 0$, $i \in N = \{1, 2, 3\}$, and $\theta \in \Omega = \{1, 2\}$.

The interpretation is the following: x_i represents the production output of agent i, θ describes consumer demand for the products. Then the first term in the goal function expresses the sales proceeds (as the price is multiplied by the output as in the Cournot oligopoly model [99]); the second term serves to describe the manufacturing costs.

To simplify the exposition, call a *pessimist* any agent believing that the demand is low ($\theta = 1$); on the other hand, an *optimist* believes the demand is high ($\theta = 2$). Let the first two agents be optimists and the third be a pessimist; suppose that all agents possess identical awareness. Evaluation of the Nash equilibrium requires solving the system of equations as follows:

$$\begin{cases} x_1^* = \dfrac{2 - x_2^* - x_3^*}{3}, \\ x_2^* = \dfrac{2 - x_1^* - x_3^*}{3}, \\ x_3^* = \dfrac{2 - x_1^* - x_2^*}{3}. \end{cases}$$

Thus, the following actions of the agents form Nash equilibrium: $x_1^* = x_2^* = 1/2$, $x_3^* = 0$.

Assume now that only the awareness of the agents changed: let agents 1 and 2 be optimists, while the third one (being a pessimist) believes all agents are also pessimists and have identical awareness. The first two agents possess identical awareness being adequately aware of the third one. What is the "equilibrium" of the agents' interaction? How can game theory predict their "stable" actions? •

To answer these questions (see Example 5.2) one has to describe and analyze mutual awareness of the agents. It requires the concepts of awareness (informational) structure and informational equilibrium (which generalizes the concept of Nash equilibrium to the case of nontrivial finite structures of agents' mutual beliefs). These concepts are described in the present chapter.

From a historical point of view, in game theory, psychology, distributed systems, and other fields of science (see surveys and references in [54, 124]) one should consider agents' *beliefs* not only about essential parameters but also about the beliefs of other agents. The set of such beliefs is called the *hierarchy of beliefs*. It may be modeled using the tree of informational structure of a reflexive game. In other words, situations of interactive decision making (modeled in game theory) require that each agent forecasts opponents' behavior prior to his or her choice. Thus, each agent should possess definite beliefs about the view of the game by his or her opponents. On the other hand, opponents should do the same. Consequently, the uncertainty regarding the game to be played generates an infinite hierarchy of beliefs of game participants.

A special case of awareness concerns *common knowledge*, when beliefs of all orders coincide. First discussions of agents' mutual awareness appeared in D. Lewis's philosophical monograph [93]. A rigorous definition of common knowledge was introduced in [6] by R. Aumann (see also the survey in [7]). Notably, common knowledge is a fact with the following properties:

1. All agents know it.
2. All agents know 1.
3. All agents know 2 and so on *ad infinitum*.

Game theory often assumes that all parameters of a game are common knowledge. Such assumption corresponds to the *objective description of a game* and enables addressing the

Nash equilibrium concept as a forecasted outcome of a noncooperative game (a game where agents do not agree about, e.g., coalitions, data exchange, joint actions, and redistribution of payoffs). Thus, the assumption regarding common knowledge allows claiming that all agents know which game they play and that their beliefs about the game coincide. (Note that in the early 1950s, when J. Nash [120] pioneered his equilibrium concept, terms such as *common knowledge* did not exist.)

Instead of an agent's action, one may consider something more complicated—an agent's *strategy*. A strategy represents a mapping of all information available to an agent into a set of his or her feasible actions. For instance, we mention strategies in a multistep game, mixed strategies, and strategies in Howard's metagames [74, 75] (see also informational extensions of games in [56]). However, in these cases the rules of play are common knowledge. Finally, it seems possible to believe that a game is chosen randomly according to a certain probability distribution making up common knowledge—the *Bayesian games*, introduced by J. Harsanyi [68]. Correspondence between Bayesian-Nash equilibrium in Bayesian games (see also the model of infinite-depth beliefs structure in [104]) and informational equilibrium in reflexive games was established in [39].

Generally, each agent may possess individual beliefs about parameters of a game. Therefore, each belief corresponds to a *subjective description of the game* [56]. Consequently, agents participate in the game, having no objective views of it or interpreting this game in different ways (e.g., rules, goals, and the roles and awareness of opponents).

The problem of designing and analyzing mathematical models of games, called *reflexive games* where agents' awareness is not common knowledge and they make decisions based on finite hierarchies of their beliefs, was solved in [40, 41]. It is worth noting that the term *reflexive games* was introduced by V. Lefebvre in 1965 [89, 91]. However, the cited work and his other publications [88, 90] represented discussions (not in the framework of game theory) of reflexion effects in interaction among subjects. Modern state of reflexive games theory is reflected in monograph [124]. The present chapter contains an introduction to the application of this theory to the problems of informational control in organizations.

5.1 Model of Informational Control

According to the approach of game theory (and mechanism design), a control problem lies in constructing a game for controlled subjects (agents) such that its outcome would be the most beneficial to a principal. Therefore, an informational control problem could be posed in the following (informal yet qualitative) way. Find a certain awareness structure such that the outcome of the reflexive game (see Section 5.2) of the agents would be the most beneficial to the principal; note that the outcome is calculated according to the concept of informational equilibrium.

Let us provide a formal statement of the control problem. Assume that the goal function of the principal, $\Phi(x, I)$, is defined on a set of real agents' actions and awareness structures. Next, suppose the principal can form any awareness structure from a certain set $\mathfrak{I}'$. Under the given awareness structure $I \in \mathfrak{I}'$, the action vector of real agents is an element of the set of equilibrium vectors $\Psi_X(I)$. We emphasize that the set $\Psi_X(I)$ may be empty in general; in the case of missed equilibrium, the principal cannot predict an outcome of the game. To avoid this problem let us introduce the set of feasible structures leading to the nonempty set of equilibria: $\mathfrak{I} = \{I \in \mathfrak{I}' \mid \Psi_X(I) \neq \varnothing\}$.

Imagine that under the specified awareness structure $I \in \mathfrak{I}$ the set of equilibrium vectors $\Psi_X(I)$ includes (at least) two elements. As a rule, one of the following assumptions is then adopted:

1. *Hypothesis of benevolence* (HB), which implies that agents always choose the equilibrium desired by the principal (in other words, preferences of the principal induce a *focal point* on the set of equilibria).
2. *Principle of maximal guaranteed result* (PMGR), that is, the principal expects the worst-case equilibrium of the game.

Using either HB or PMGR, one has the *problem of informational control* in two settings as follows:

$$\max_{x \in \Psi_X(I)} \Phi(x, I) \xrightarrow[I \in \Im]{} \max; \tag{5.1}$$

$$\min_{x \in \Psi_X(I)} \Phi(x, I) \xrightarrow[I \in \Im]{} \max. \tag{5.2}$$

Naturally, if for any $I \in \Im$ the set $\Psi_X(I)$ consists of a single element, formulas (5.1) and (5.2) coincide.

In the sequel, the problem (5.1) (alternatively, (5.2)) will be called an informational control problem in the form of the goal function.

Now, give an alternative formulation to the problem of informational control (being independent of the goal function of the principal). Assume that the principal wants the agents to choose an action vector $x \in X'$. The question arises: For which vectors and by which awareness structure I would the principal achieve this? In other words, the second possible formulation of the informational control problem is to find the following components. First, the attainability set, viz. the one composed of the vectors $x \in X'$ such that for each of them the set of awareness structures $\Psi_I(x) \cap \Im$

$$\text{is nonempty} \tag{5.3}$$

or

$$\text{consists of a single element.} \tag{5.4}$$

Second, the corresponding feasible awareness structures $I \in \Psi_I(x) \cap \Im$ for each vector x meeting the aforementioned property. Note that condition (5.3) "corresponds" to hypothesis of benevolence, while (5.4) "corresponds" to the principle of the maximal guaranteed result (see Section 1.2).

Problem (5.3) (alternatively, (5.4)) will be referred to as the problem of informational control in the form of the attainability set. Once again, it should be underlined that the second formulation of the problem does not depend on the goal function of the principal; it merely reflects the possibility of bringing the system to a certain state by informational control.

In both formulations of the problem the principal may either be interested in stability of the resulting informational equilibrium or not (see Section 5.3 for the corresponding issues). A stable informational equilibrium being required (i.e., when the system has to be rendered stable), one should substitute Ψ^s for Ψ in the previous formulas. In addition, the term *equilibrium* should be replaced with a *stable equilibrium.*

The Model of Informational Control

The suggested model of informational control is illustrated in Figure 5.1. It includes an agent (or several agents) and a principal. Each agent is characterized by the cycle "awareness of the agent → action of the agent → result observed by the agent → awareness of the agent"; generally speaking, these components vary for different agents. At the same time, the cycle could be viewed common for the whole controlled subsystem (i.e., for the complete set of the agents). This feature is indicated by the word "*Agent(s)*" in Figure 5.1.

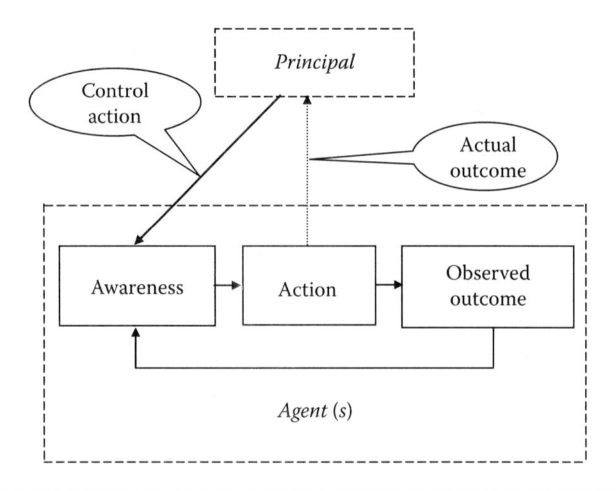

Figure 5.1 The model of informational control.

The interaction between the agent (agents) and the principal is characterized by the following elements:

1. An informational impact of the principal, resulting in formation of a certain awareness of the agent (agents)*
2. An actual outcome of the agent's action (or agents' actions), which has an impact on preferences of the principal

Let us discuss the model presented by Figure 5.1 in detail.

The mathematical framework used to model game-theoretic interaction of the agents is provided by *reflexive games*, where agents choose actions based on their *awareness structures*: for example, hierarchical beliefs about essential parameters of the situation ("state of nature") and beliefs about beliefs of the opponents (other agents). Therefore, in terms of reflexive games the agent's awareness is modeled by his or her

* Note that it is possible to study the influence of a principal on the outcome observed by an agent (agents); see the chain "principal → observed outcome" in Figure 5.1. However, such investigation lies beyond the scope of this textbook (in a certain sense, it reduces the differences between informational control and motivational control).

awareness structure. (The awareness of the whole controlled subsystem is just a union of agents' awareness structures.)

An agent chooses an action based on his or her awareness structure. Under the given awareness structure, actions of the agents are, in fact, the components of an *informational equilibrium*, which is a solution of the reflexive game. The notion of informational equilibrium generalizes the concept of Nash equilibrium (probably the most popular solution concept for noncooperative games).

In many cases agents' awareness about the situation and about beliefs of opponents may be inadequate. Hence, the result of a reflexive game observed by an agent either meets his or her expectations or not. This is defined by the following factors:

1. How adequate is agent's awareness at the moment of choosing the action?
2. How complete is the information observed by the agent about the outcome of the game?

For instance, the observed outcome could be the value of his or her goal function, actions of the opponents, or the true value of an uncertain parameter (state of nature). Generally, the agent observes the value of some function, which depends on the state of nature and actions of the opponents. This function is known as the *observation function*, and its impact on the awareness is illustrated by the chain "observed action → awareness." Imagine that all agents observe the result they really reckon on (notably, for every agent the actual value of the observation function coincides with the expected one). In this case, the assumption regarding fixed (invariable) awareness structure seems natural, and the informational equilibrium appears *stable*.

Now consider interaction between the agents and the principal. Implementing informational control, the principal (as usual) strives to maximize his or her utility. Assume that the principal can form any awareness structure from a certain feasible set. The problem of informational control may be posed as follows. Find an awareness structure from the set of feasible structures to maximize principal's goal function in the corresponding informational equilibrium (perhaps taking into account the principal's costs to form this awareness structure).

Let us underline an important aspect. Within the proposed model, we adopt the assumption that the principal may form *any* awareness structure of the agents. The issue of how the principal should "convince" the agents that specific states of nature and beliefs of the opponents take place is not discussed here.

However, within the current model one may classify the methods of control action applied to the agents' awareness (to form a certain structure). These methods include the following:

1. *Informational regulation*, representing a purposeful impact on information about the state of nature
2. *Reflexive control*, representing a purposeful impact on information about beliefs of the opponents
3. *Active forecast*, representing a purposeful reporting of information on future values of the parameters that depend on the state of nature and actions of the agents

The Problems of Informational Control: A Classification

The present chapter focuses on a two-level OS with a single principal and several agents under incomplete awareness of the agents; that is, each subject may have individual beliefs about the state of nature or the types of his or her opponents.

We will consider the problem of informational control:

1. In the form of the goal function or in the form of the attainability set

2. Involving the hypothesis of benevolence (HB) or the principle of the maximal guaranteed result (PMGR)
3. With or without the stability requirement

The choice of a specific setting (there are eight altogether) is subject to the actual situation being modeled. In addition, it is necessary to establish a relation between the awareness structure and the action vector of the agents (to perform informational stability analysis); the experience indicates this stage is the most complicated and time-consuming for a researcher.

5.2 Reflexive Games

Consider the set of agents: $N = \{1, 2,\ldots, n\}$. Denote by $\theta \in \Omega$ the uncertain parameter (we believe that the set Ω is common knowledge for all agents). The *awareness structure* I_i of agent i includes the following elements. First, the belief of agent i about the parameter θ; denote it by θ_i, $\theta_i \in \Omega$. Second, the beliefs of agent i about the beliefs of the other agents about the parameter θ; denote them by θ_{ij}, $\theta_{ij} \in \Omega$, $j \in N$. Third, the beliefs of agent i about the beliefs of agent j about the belief of agent k; denote them by θ_{ijk}, $\theta_{ijk} \in \Omega$, $j, k \in N$. (This process is generally infinite.) Note in the sequel we employ the terms of *informational structure* and *hierarchy of beliefs* as synonyms for the awareness structure.

Therefore, the awareness structure I_i of agent i is specified by the set of values $\theta_{ij_1\ldots j_l}$, where l runs over the set of nonnegative integer numbers, $j_1, \ldots, j_l \in N$, while $\theta_{i_1\ldots i_l} \in \Omega$.

The *awareness structure I of the whole game* is defined in a similar manner; in particular, the set of the values $\theta_{i_1\ldots i_l}$ is employed, with l running over the set of nonnegative integer numbers, $j_1, \ldots, j_l \in N$, and $\theta_{ij_1\ldots j_l} \in \Omega$. We emphasize that the agents are not aware of the whole structure I; each of them knows only a substructure I_i.

Thus, an awareness structure is an infinite n-tree; the corresponding nodes of the tree describe specific awareness of real agents from the set N, and also phantom agents (complex reflexions of real agents in the mind of their opponents).

A reflexive game Γ_I is a game defined by the following tuple [40, 41]:

$$\Gamma_I = \{N, (X_i)_{i \in N}, f_i(\cdot)_{i \in N}, \Omega, I\}, \tag{5.5}$$

where N stands for a set of real agents, X_i is a set of feasible actions of agent i, $f_i(\cdot)$: $\Omega \times X' \to \Re^1$ is his or her goal function ($i \in N$), Ω indicates a set of feasible values of the uncertain parameter, and I designates the awareness structure.

Therefore, a reflexive game generalizes the notion of a normal-form game (determined by the tuple $\{N, (X_i)_{i \in N}, f_i(\cdot)_{i \in N}\}$) to the case when the agents' awareness is reflected by a hierarchy of their beliefs (i.e., the informational structure I). Within the framework of the accepted definition, a "classical" normal-form game is a special case of a reflexive game (a game under common knowledge among the agents). Consider the "extreme" case when the state of nature appears to be common knowledge; for a reflexive game, the solution concept (proposed in the present textbook based on informational equilibrium) turns out equivalent to the Nash equilibrium concept.

The set of relations among the elements characterizing the agents' awareness may be illustrated by a tree (see Figure 5.2). Note that the awareness structure of agent i is represented by the corresponding subtree starting from the node θ_i.

Let us make the following important remark. Here and in the sequel we confine ourselves to the consideration of "point-type" awareness structures (the components consist of the elements belonging to the set Ω). More general models (e.g., an interval uncertainty, a probabilistic uncertainty, or a fuzzy uncertainty) are not considered.

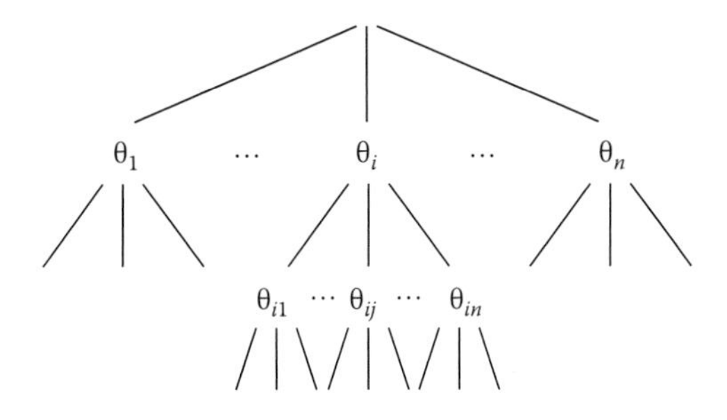

Figure 5.2 Tree of informational structure.

Strategic and Informational Reflexion

Summarizing the previous statements, we underline that in a reflexive game the agents' awareness is not common knowledge. Following the concepts of game theory and the reflexive models of decision making, one should distinguish between strategic reflexion and informational reflexion [124].

Informational reflexion is a process and result of agent's thinking of, first, the values of uncertain parameters and, second, what knowledge and beliefs about these values his or her opponents (other players) actually possess. In fact, a "game" component in informational reflexion does not occur since a player makes no decisions.

In other words, informational reflexion covers the agent's awareness about the nature of reality (the actual game) and the reflexive reality (the game being seen by the others). Heuristically, informational reflexion foreruns strategic reflexion, as decisions are based on certain information. *Strategic reflexion* is a process and result of agent's thinking of what decision-making principles are used by his or her opponents (under the awareness being assigned to them by this agent as the result of informational reflexion). Therefore, informational reflexion occurs only under incomplete information, and the result of such reflexion is used in the process of action selection. At the same time, strategic reflexion is even intrinsic to the case of complete information, anticipating the agent's decision regarding the choice of a specific action (a strategy). In other words, informational reflexion and strategic reflexion may be studied independently; however, both take place in the conditions of incomplete information.

To proceed and formulate a series of definitions and properties, we introduce the following notation [40, 41]:

Σ_+ stands for a set of finite sequences of indexes belonging to N;
Σ is the sum of Σ_+ and the empty sequence;
$|\sigma|$ indicates the number of indexes in the sequence $\sigma \in \Sigma$ (for the empty sequence, it equals zero); this parameter is known as the length of an index sequence.

Imagine that θ_i represents the belief of agent i about the uncertain parameter, while θ_{ii} means the belief of agent i about his or her own belief. It seems then natural that $\theta_{ii} = \theta_i$. In other words, agent i is well-informed on his or her own beliefs. Moreover, he or she assumes that the remaining agents possess the same property. Formally, this means that the *axiom of self-awareness* [40, 41] is satisfied:

$$\forall\ i \in N\ \forall\ \tau, \sigma \in \Sigma\ \theta_{\tau ii\sigma} = \theta_{\tau i\sigma}.$$

In the sequel we suppose that the axiom of self-awareness holds; in particular, this axiom implies the following. Being aware of θ_τ for all $\tau \in \Sigma_+$ such that $|\tau| = \gamma$, an agent may explicitly evaluate θ_τ for all $\tau \in \Sigma_+$ with $|\tau| < \gamma$.

In addition to the awareness structures I_i ($i \in N$), the researcher may also analyze the awareness structures I_{ij} (i.e., the awareness of agent j according to the belief of agent i), I_{ijk}, and so on. Let us identify the awareness structure with the agent being characterized by it. In this case, one may claim that n *real* agents (*i-agents*, where $i \in N$) having the awareness structures I_i also play with *phantom agents*

(τ-*agents*, where $\tau \in \Sigma_+$, $|\tau| \geq 2$) having the awareness structures $I_\tau = \{\theta_{\tau\sigma}\}$, $\sigma \in \Sigma$. It should be emphasized that phantom agents exist merely in the minds of real agents; still, they have an impact on their actions; these aspects will be discussed.

We now introduce *identical awareness structures*, the basic notion for this chapter.

Awareness structures I_λ and I_μ (λ, $\mu \in \Sigma_+$) are said to be *identical* if the following conditions are met:

1. $\theta_{\lambda\sigma} = \theta_{\mu\sigma}$ for any $\sigma \in \Sigma$.
2. The last indexes in the sequences λ and μ coincide.

Identity of these structures will be designated by $I_\lambda = I_\mu$.

The first condition in the definition is transparent, but the second one should be discussed in greater detail. The matter is that in the sequel we consider actions of a τ-agent depending on his or her awareness structure I_τ and his or her goal function f_i (note the latter is determined by the last index in the sequence τ). Therefore, it appears convenient for us to believe that identity of the awareness structures means identity of the goal functions.

A λ-agent is said to be τ-subjectively *adequately aware* of the beliefs of a μ-agent (or, in a compact form, of a μ-agent) if and only if

$$I_{\tau\lambda\mu} = I_{\tau\mu} \ (\lambda,\ \mu \in \Sigma_+,\ \tau \in \Sigma).$$

We will use the expression $I_\lambda >_\tau I_\mu$ to indicate the τ-subjective adequate awareness of the λ-agent about the μ-agent.

The notion of identical awareness structures allows for introducing another relevant property—the complexity of the structure. We underline that (along with the structure I) there exists a denumerable set composed of the awareness structures I_τ, $\tau \in \Sigma_+$ such that among them one may separate out certain classes of pairwise nonidentical structures by the identity relation. The number of the mentioned classes is naturally referred to as the *complexity of the awareness structure.*

We will say that the awareness structure I has *finite complexity* $\nu = \nu(I)$, if there is a finite set of pairwise nonidentical structures $\{I_{\tau_1}, I_{\tau_2}, \ldots, I_{\tau_\nu}\}$, $\tau_l \in \Sigma_+, l \in \{1, \ldots, \nu\}$ such that any structure I_σ $\sigma \in \Sigma_+$, has an identical structure from this set. Otherwise, the structure I possesses the infinite complexity: $\nu(I) = \infty$.

A finite-complexity awareness structure is called *finite* (however, the corresponding awareness tree is infinite). If this is not the case, the awareness structure is said to be *infinite.*

Obviously, the minimum possible complexity of an awareness structure equals the number of real agents participating in the game (one can check that awareness structures for real agents are pairwise nonidentical by definition).

Any (finite or denumerable) set of pairwise nonidentical structures I_τ, $t \in S_+$ such that any structure I_σ, $\sigma \in \Sigma_+$ is identical to one of them, is referred to as a *basis* of the awareness structure I.

Suppose the awareness structure I has finite complexity; then it is possible to estimate the maximum length of the index sequence γ such that, given all structures $I_\tau, \tau \in \Sigma_+$, $|\tau| = \gamma$, one can find the remaining structures. In a certain sense, this length characterizes *the rank of reflexion* necessary to describe the awareness structure.

We will say that the awareness structure I, $\nu(I) < \infty$, has *finite depth* $\gamma = \gamma\ (I)$ when the following conditions hold:

1. For any structure I_σ, $\sigma \in \Sigma_+$, there exists an identical structure I_τ, $\tau \in \Sigma_+$, $|\tau| \le \gamma$.
2. For any integer positive number ξ, $\xi < \gamma$ there exists a structure I_σ, $\sigma \in \Sigma_+$, being identical to none of the structures I_τ, $\tau \in \Sigma_+$, $|\tau| = \xi$.

If $\nu(I) = \infty$, the depth is also considered infinite: $\gamma(I) = \infty$.

The notions of complexity and depth of awareness structures could be introduced in the τ-subjective context of the game, as well. In particular, the depth of an awareness structure of the game in the view of a τ-agent, $\tau \in \Sigma_+$, will be called the *reflexion rank* of the τ-agent.

Graph of the Reflexive Game

If an awareness structure has a finite complexity, one may draw a *graph of the reflexive game* [40, 41, 124]; it illustrates interconnection among actions of real and phantom agents.

The nodes of such directed graphs are represented by the actions x_t, $t \in S_+$ that correspond to the pairwise nonidentical awareness structures I_t (alternatively, by the components of the awareness structure θ_τ or simply by the number τ of a certain real or phantom agent, $t \in S_+$).

The nodes are connected by means of arcs according to the following rule. Each node $x_{\sigma i}$ has arcs coming from $(n - 1)$ nodes that correspond to the structures $I_{\sigma ij}$, $j \in N \setminus \{i\}$. Two nodes being connected by two opposite arcs, the edge is represented by two arrows.

Let us emphasize that the graph of reflexive game meets the system of equations (5.6) (i.e., it satisfies the definition of informational equilibrium); nevertheless, it is possible that the solution to this system does not exist.

Thus, the graph G_I of the reflexive game Γ_I (see the notion of reflexive game above) having a finite complexity of the awareness structure is defined in the following way:

1. Nodes of the graph G_I correspond to real and phantom agents participating in the reflexive game (in other words, to the pairwise nonidentical awareness structures);
2. Arcs of the graph G_I describe mutual awareness of the agents; that is, if there exists a path from a certain (real or phantom) agent to another one, the latter is adequately aware of the former.

Suppose that nodes of the graph G_I describe beliefs of the corresponding agent about the state of nature. Then the reflexive game Γ_I with the finite awareness structure I may be specified by the tuple $\Gamma_I = \{N, (X_i)_{i \in N}, f_i(\cdot)_{i \in N}, G_I\}$, where N is a set of real agents, X_i stands for a set of feasible actions of agent i, $f_i(\cdot)$: $\Omega \times X' \to R^1$ gives his or her goal function, $i \in N$, and G_I designates the graph of the reflexive game.

In many cases it is convenient (and even visual) to describe a reflexive game exactly in terms of the graph G_I (thus, not involving the tree of informational structure). Several examples of such graphs are given in the following section.

5.3 Informational Equilibrium

Assume that the awareness structure I of a game is given; this means that the awareness structures are also defined for all (real and phantom) agents. Within the framework of the hypothesis of rational behavior, the choice of an action x_t performed by a t-agent is described by his or her awareness structure I_t. Hence, the mentioned structure being available, one may model an agent's reasoning and evaluate his or her action. On the other hand, while choosing his or her action, the agent models actions of the remaining agents (i.e., performs reflexion). Therefore, estimating the game outcome, we should account for the actions of real and phantom agents.

The set of the actions x_t^*, $t \in \Sigma_+$, is called an *informational equilibrium* [40, 41, 124], if the following conditions are met:

(1) the awareness structure I possesses finite complexity ν;

(2) $\forall \lambda, \mu \in \Sigma$, $I_{\lambda_i} = I_{\mu_i} \Rightarrow x^*_{\lambda_i} = x^*_{\mu_i}$;

(3) $\forall\ i \in N, \forall\ \sigma \in \Sigma$,

$$x^*_{\sigma_i} \in \operatorname*{Arg\,max}_{x_i \in X_i} f_i(\theta_{\sigma_i}, x^*_{\sigma_i,1}, \ldots, x^*_{\sigma_i,i-1}, x_i, x^*_{\sigma_i,i+1}, \ldots, x^*_{\sigma_i,n}). \tag{5.6}$$

Condition 1 in the definition of the informational equilibrium claims that the reflexive game involves a finite number of real and phantom agents. Condition 2 expresses, in fact, the requirement that the agents with identical awareness choose identical actions. Condition 3 reflects rational behavior of the agents: each agent strives for maximizing the individual goal function via a proper choice of his or her action. For this, the agent substitutes actions of the opponents into his or her goal function; the actions are rational in the view of the considered agent (according to the available beliefs of the remaining agents).

It would seem that condition 2 makes it necessary to solve an infinite (denumerable) number of equations (yielding infinitely many values x_t^*) to evaluate an informational equilibrium. However, it turns out that the actual number of equations and solutions is finite.

Assertion 5.1 [40, 41, 124]

Suppose there exists an informational equilibrium x_t^*, $\tau \in \Sigma_+$. Then it consists of (at most) υ pairwise different actions, and the system (5.6) includes at most υ pairwise different equations.

Proof.

Let x_t^*, $\tau \in \Sigma_+$, be an informational equilibrium. Then finiteness of the awareness structure and condition 2 immediately imply that the number of pairwise different values x_τ^* constitutes (at most) n.

Now, consider two arbitrary identical awareness structures: $I_\lambda = I_\mu$. In this case, we have $\theta_\lambda = \theta_\mu$ and $x_\lambda^* = x_\mu^*$. Next, for any $i \in N$ the equality $I_{\lambda i} = I_{\mu I}$ holds; hence, $x_{\lambda i}^* = x_{\mu i}^*$. This means that two equations coincide in the system (5.6) with the actions x_λ^* and x_μ^* in the left-hand side. Since there are υ pairwise different awareness structures, the number of pairwise different conditions (5.6) does not exceed υ. •

Therefore, to find an informational equilibrium x_t^*, $\tau \in \Sigma_+$, one must write down υ conditions of the form (5.6) for each of

υ pairwise different values x_τ^* that correspond to the pairwise different awareness structures I_τ.

All agents being identically aware of the situation, complexity of the awareness structure is minimal and equal to the number of the agents. In this case, the system (5.6) gives the definition of Nash equilibrium, while the informational equilibrium becomes a Nash equilibrium.

Let us summarize the results. When real agents have identical awareness (i.e., reflexive reality is common knowledge), an informational equilibrium turns into a Nash equilibrium (no phantom agents "arise").

The informational equilibrium (see (5.6)) is somewhat cumbersome; one would often hardly observe the relation between the informational structure and the informational equilibrium. The previous graph of reflexive game represents, first, a convenient description for mutual awareness of agents and, second, an efficient tool for the analysis of the properties of informational equilibrium.

We provide several examples of informational equilibrium evaluation using the graph of reflexive game. Examples 5.1 and 5.2 differ only in the awareness structures. According to condition 2 of the notion of informational equilibrium, in Example 5.1 similar properties hold for equilibrium actions x_σ^*. Obviously, any awareness structure appears identical to one of those structures forming the basis $\{I_1, I_2, I_3\}$. Hence, complexity of the given awareness structure in Example 5.1 equals 3, while the depth makes 1. The corresponding graph of the reflexive game is illustrated in Figure 5.3.

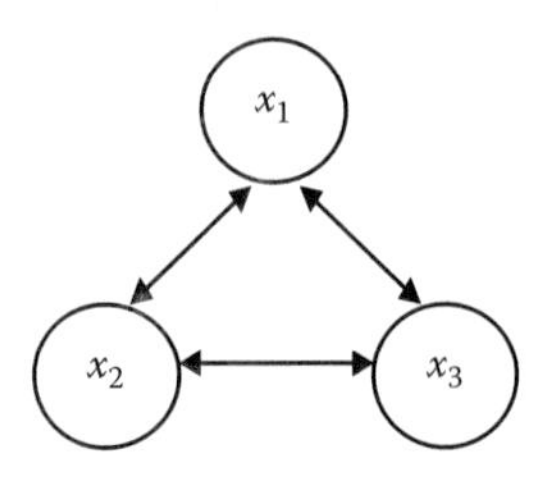

Figure 5.3 The graph of the reflexive game in Example 5.1.

Example 5.2

Assume (as in the introduction to the present chapter) that agents 1 and 2 are optimists and that the third one (being a pessimist) believes all agents are also pessimists and have identical awareness. The first two agents possess identical awareness being adequately aware of the third one.

The complexity and depth of the stated awareness structure constitute 5 and 3, respectively. Figure 5.4 shows the graph of the reflexive game.

The following system of equations should be solved to find the informational equilibrium (see (5.6)):

$$\begin{cases} x_1^* = \dfrac{2 - x_2^* - x_3^*}{3} \\ x_2^* = \dfrac{2 - x_1^* - x_3^*}{3} \\ x_3^* = \dfrac{2 - x_{31}^* - x_{32}^*}{3} \\ x_{31}^* = \dfrac{1 - x_{32}^* - x_3^*}{3} \\ x_{32}^* = \dfrac{1 - x_{31}^* - x_3^*}{3} \end{cases} \Leftrightarrow \begin{cases} x_1^* = \dfrac{9}{20}, \\ x_2^* = \dfrac{9}{20}, \\ x_3^* = \dfrac{1}{5}, \\ x_{31}^* = \dfrac{1}{5}, \\ x_{32}^* = \dfrac{1}{5}. \end{cases}$$

Consequently, the actions of real agents in the informational equilibrium are, in fact,

$$x_1^* = x_2^* = 9/20,\ x_3^* = 1/5.$$

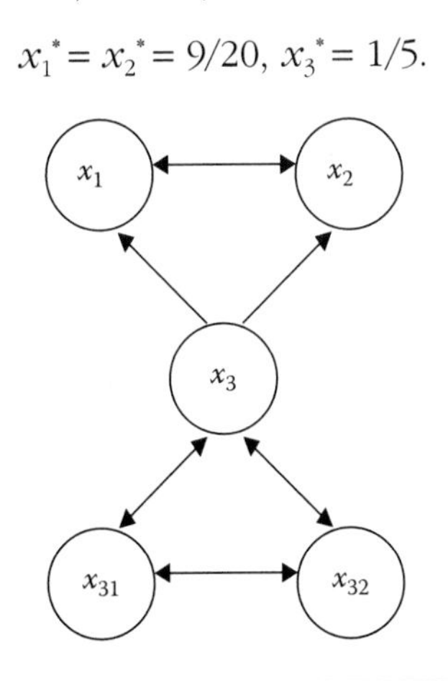

Figure 5.4 The graph of the reflexive game in Example 5.2.

Note that changing only the awareness (i.e., passing from the awareness structure of Figure 5.3 to that of Figure 5.4) allows for increasing the total production output of the agents and their total surplus! •

Stable Informational Equilibrium [42]

The classical concept of Nash equilibrium is remarkable for its self-sustained nature. Notably, assume that a repetitive game takes place and all agents (except agent i) choose the same equilibrium actions. Then agent i benefits nothing by deviating from his or her equilibrium action; evidently, this feature is directly related to the following. Beliefs of all agents about reality are adequate; that is, the state of nature appears common knowledge.

Generally speaking, the situation may change in the case of the informational equilibrium. Indeed, after a single play of the game some agents (or even all of them) may observe an unexpected outcome due to an inconsistent belief about the state of nature (or due to inadequate awareness of opponents' beliefs). In addition, the self-sustained nature of the equilibrium is violated; actions of agents may change as the game is repeated.

However, in some cases a self-sustained equilibrium takes place for differing (generally, incorrect) beliefs of the agents. As a matter of fact, such a situation occurs when each agent (real or phantom) observes the game outcome he or she expects. To develop a formal framework, we need to refine the definition of reflexive game.

Recall a reflexive game is defined by the tuple $\{N, (X_i)_{i \in N}, f_i(\cdot)_{i \in N}, \Omega, I\}$, where $N = \{1, 2, \ldots, n\}$ means a set of game participants (players, agents), X_i indicates a set of feasible actions of agent i, $f_i(\cdot)$: $\Omega \times X' \rightarrow \Re^1$ represents his or her goal function, $i \in N$, and I is the awareness structure. Let us augment this structure with a set of functions $w_i(\cdot)$: $\Omega \times X' \rightarrow W_i$, $i \in N$, each mapping the vector (θ, x) into an element w_i of a certain set W_i. The element w_i is exactly what agent i observes as the outcome of the game.

In the sequel the function $w_i(\cdot)$ will be referred to as the *observation function* of agent i. Suppose that the observation functions are common knowledge of the agents.

Imagine that $w_i(\theta, x) = (\theta, x)$, i.e., $W_i = \Omega \times X'$; then agent i observes both the state of nature and the actions of all agents. On the contrary, the set W_i being composed of a single element, agent i observes nothing.

Suppose the reflexive game admits an informational equilibrium x_t, $\tau \in \Sigma_+$ (recall t is an arbitrary nonempty sequence of indexes belonging to N). Next, fix $i \in N$ and consider agent i. He or she expects to observe the following outcome of the game:

$$w_i\,(q_i, x_{i1}, \ldots, x_{i,\,i-1}, x_i, x_{i,\,i+1}, \ldots, x_{in}). \tag{5.7}$$

Actually, he or she observes

$$w_i\,(q, x_1, \ldots, x_{i-1}, x_i, x_{i+1}, \ldots, x_n). \tag{5.8}$$

Therefore, the stability requirement for agent i implies coincidence of the values (5.7) and (5.8) (again, we indicate these are the elements of a certain set W_i).

Assume the values (5.7) and (5.8) are equal to each other; in other words, agent i has no doubts regarding validity of his or her beliefs after the game. Meanwhile, is it enough for the agent to choose the same action x_i in the next game? Clearly, the answer is negative; the example below gives an illustration.

Example 5.3

Consider a reflexive bimatrix game, where $\Omega = \{1, 2\}$. The gains are specified by the bimatrices, see Figure 5.5

$$\theta = 1 \qquad\qquad \theta = 2$$

$$\begin{pmatrix} (1,1) & (0,0) \\ (0,1) & (2,0) \end{pmatrix} \quad \begin{pmatrix} (0,1) & (1,2) \\ (1,1) & (2,2) \end{pmatrix}$$

Figure 5.5 The matrices of players' gains in Example 5.3.

(agent 1 chooses the row, while agent 2 chooses the column: $X_1 = X_2 = \{1;\ 2\}$).

The graph of the corresponding reflexive game is shown in Figure 5.6.

Moreover, set $\theta = \theta_1 = 1$, $\theta_2 = \theta_{21} = 2$ and suppose each agent observes his or her gain (i.e., for any agent the observation function coincides with his or her gain function). Obviously, the informational equilibrium is provided by the combination $x_1 = x_2 = x_{21} = 2$; that is, agents 1 and 2 (as well as 21-agent and the remaining phantom agents) choose the second actions. However, after the game the real state of nature ($\theta = 1$) becomes known to agent 2; note the latter gains 0 instead of the expected value of 2. Hence, next time agent 2 would choose the action $x_2 = 1$, motivating agent 1 to change his or her action (i.e., to choose $x_1 = 1$). •

Therefore, a stable equilibrium requires the phantom ij-agent ($i, j \in N$) also to observe the "necessary" value. As the result of the game, the agent in question expects to observe

$$w_j\,(\theta_{ij}, x_{ij1}, \ldots, x_{ij,\,j\text{-}1}, x_{ij}, x_{ij,\,j+1}, \ldots, x_{ijn}). \tag{5.9}$$

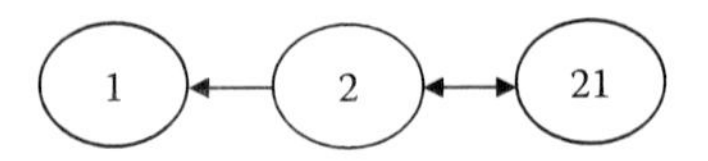

Figure 5.6 The graph of reflexive game in Example 5.3.

Actually (in other words, *i*-subjectively, since *ij*-agent exists in the mind of *i*-agent) he or she observes the value

$$w_j\,(\theta_i,\ x_{i1},\ \ldots,\ x_{i,\,j-1},\ x_{ij},\ x_{i,\,j+1},\ \ldots,\ x_{in}). \qquad (5.10)$$

Consequently, for *ij*-agent the stability requirement implies coincidence between (5.9) and (5.10).

In general case (i.e., for τi-agent, $\tau i \in \Sigma_+$), we introduce the following definition of stability.

An *informational equilibrium* $x_{\tau i,}$ $\tau i \in \Sigma_+$, is said to be *stable* under a given awareness structure I if for any $\tau i \in \Sigma_+$ the following equality holds:

$$w_i\,(\theta_{\tau i},\ x_{\tau i1},\ \ldots,\ x_{\tau i,\,i-1},\ x_{\tau i},\ x_{\tau i,\,i+1},\ \ldots,\ x_{\tau in}) \qquad (5.11)$$

$$= w_i\,(\theta_\tau,\ x_{\tau 1},\ \ldots,\ x_{\tau,\,i-1},\ x_{\tau i},\ x_{\tau,\,i+1},\ \ldots,\ x_{\tau n}).$$

If an informational equilibrium is not stable in the mentioned sense, we will call it *unstable*. For instance, the informational equilibrium in Example 5.3 appears unstable.

Assertion 5.2 [42]

Consider an informational structure I of complexity υ. Suppose there exists an informational equilibrium $x_{\tau i}$, $ti \in \Sigma_+$. Then the system (5.11) includes at most n pairwise different conditions.

Proof.

Consider two (arbitrary) identical awareness structures: $I_{\lambda i} = I_{\mu i}$. Since x_{ti} is an equilibrium, we have $\theta_{\lambda i} = \theta_{\mu i}$, $x_{\lambda i} = x_{\mu i}$, $I_{\lambda ij} = I_{\mu ij}$, and $x_{\lambda ij} = x_{\mu ij}$ for any $j \in N$. Thus, the stability conditions (5.11) are identical for λi- and μi-agents. There are υ pairwise different awareness structures; hence, the number of pairwise different conditions (5.11) does not exceed υ. •

True and False Informational Equilibria

We will divide stable informational equilibria into two classes, viz. *true* and *false* equilibria. Let us give an example.

Example 5.4

Consider a game of three agents with the goal functions

$$f_i(r_1, x_1, x_2, x_3) = x_i - \frac{x_i(x_1 + x_2 + x_3)}{r_i}$$

where $x_i \geq 0$, $i \in N = \{1, 2, 3\}$. The goal functions are common knowledge (except the *types* of the agents – their private parameters $r_i > 0$). The vector $r = (r_1, r_2, r_3)$ composed of the types may be treated as the state of nature. It is supposed that every agent knows his or her own type.

The graph of the reflexive game is shown in Figure 5.7; note that $r_2 = r_3 = r$, $r_{21} = r_{23} = r_{31} = r_{32} = c$. Each agent knows his or her type and observes the sum of the opponents' actions.

Obviously, this game has the informational equilibrium

$$\begin{aligned} x_2 &= x_3 = (3r - 2c)/4, \\ x_{21} &= x_{23} = x_{31} = x_{32} = (2c - r)/4, \\ x_1 &= (2r_1 - 3r + 2c)/4. \end{aligned} \tag{5.12}$$

The stability conditions (see (5.11)) then yield

$$x_{21} + x_{23} = x_1 + x_3,\ x_{31} + x_{32} = x_1 + x_2. \tag{5.13}$$

We have formulated the conditions for 2 and 3 agents only because their counterparts for 1, 21, 23, 31, and 32 agents are trivial.

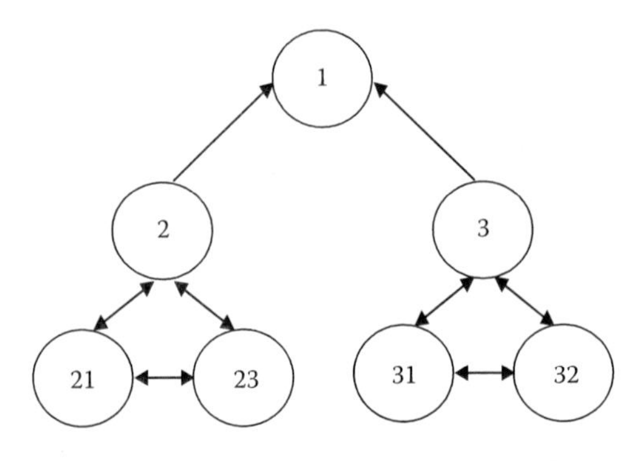

Figure 5.7 The graph of the reflexive game in Example 5.4.

Substitute (5.12) into (5.13) to obtain the following equality as a necessary and sufficient condition of stability:

$$2c = r_1 + r. \qquad (5.14)$$

Let the condition (5.14) be met. Then equilibrium actions of real agents are determined by

$$x_2 = x_3 = (3r - r_1)/4,\ x_1 = (3r_1 - 2r)/4. \qquad (5.15)$$

Now, assume that the types of the agents are common knowledge (see Figure 5.8).

One would easily ascertain that, in the case of common knowledge, the unique equilibrium is specified by (5.15). •

Therefore, the condition (5.14) being satisfied leads to a counterintuitive situation. Agents 2 and 3 have false beliefs (Figure 5.7); nevertheless, their equilibrium actions (5.15) are exactly the same as under the identical awareness (Figure 5.8). Call such stable equilibrium true.

Suppose the set of actions $x_{\tau i}$, $\tau i \in \Sigma_+$, represents a stable informational equilibrium. It will be referred to as a *true* equilibrium if the set $(x_1, \ldots, x_n)$ is an equilibrium under common knowledge about the state of nature θ (or about the set of the agents' types $(r_1, \ldots, r_n)$).

In particular, this definition implies that under common knowledge any informational equilibrium is true. Let us show another case when any informational equilibrium is also true.

Assertion 5.3 [42]

Let the goal functions of the agents be defined by

$$f_i\,(r_i, x_1, \ldots, x_n) = \varphi_i\,(r_i, x_i, y_i(x_{-i})),$$

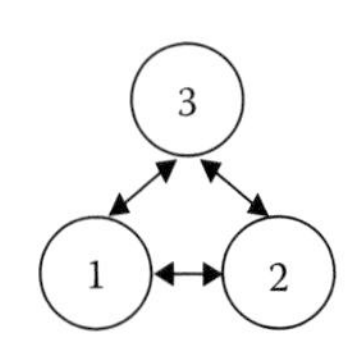

Figure 5.8 The common knowledge in Example 5.4.

while the observation functions are $w_i(\theta, x) = y_i(x_{-i})$, $i \in N$; the corresponding interpretation is as follows. The gain of every agent depends on his or her type, action and observation function (which depends on actions of the opponents, but not on their types). Then any stable equilibrium is true.

Proof.

Assume that $x_{\tau i}$, $\tau i \in \Sigma_+$ is a stable informational equilibrium and the conditions of Assertion 5.3 hold. Then for any $i \in N$ we obtain

$$x_i \in \operatorname{Arg}\max_{y_i \in X_i} f_i(r_i, y_i, x_{i,-i}) = \operatorname{Arg}\max_{y_i \in X_i} \varphi_i(r_i, y_i, y_i(x_{i,-i})).$$

Due to the stability property, the equality

$$y_i(x_{i,-i}) = y_i(x_{-i})$$

takes place, resulting in

$$x_i \in \operatorname{Arg}\max_{y_i \in X_i} \varphi_i(r_i, y_i, y_i(x_{-i})) = \operatorname{Arg}\max_{y_i \in X_i} f_i(r_i, y_i, x_{-i}).$$

Since $i \in N$ is arbitrary, the last formula means that the set $(x_1, \ldots, x_n)$ gives an equilibrium under complete information. •

A stable informational equilibrium that is not true in the above sense is said to be a *false* equilibrium.

In other words, a false equilibrium is a stable informational equilibrium, which is not the equilibrium in the case of identical awareness of the agents (under common knowledge).

Example 5.5

Consider the reflexive bimatrix game with $\Omega = \{1, 2\}$. The gains are specified by the bimatrices (see Figure 5.9) (agent 1 chooses the corresponding row, while agent 2 chooses the column: $X_1 = X_2 = \{1; 2\}$).

Next, let the actual situation be $q = 2$, while both agents consider $\theta = 1$ as common knowledge. Each agent observes

$$\begin{array}{cc} \theta = 1 & \theta = 2 \\ \begin{pmatrix} (2,2) & (4,1) \\ (1,4) & (3,3) \end{pmatrix} & \begin{pmatrix} (2,2) & (0,3) \\ (3,0) & (1,1) \end{pmatrix} \end{array}$$

Figure 5.9 The matrices of players' gains in Example 5.5.

the pair (x_1, x_2), which provides the corresponding observation function.

In the informational equilibrium both agents choose 1. Imagine the actual state of nature is common knowledge; then each agent would choose action equal to 2. Therefore, the gains of the agents in the informational equilibrium are greater in comparison to the case when the actual state of nature is common knowledge. •

Informational Impact

In this subsection we discuss some ways for the principal to exert an informational impact on the agents (with the aim of forming a specific awareness structure). In particular, these ways include informational regulation, reflexive control, and active forecasting.

In the context of the previous informational control model, this subsection corresponds to the chain "the principal → awareness of the agent (agents)" (see Figure 5.1). Being aware of the limitations of mathematical modeling of human behavior (especially, that of the game-theoretic approach in informational control), let us consider possible types of informational impact.

The principal can send the following classification of messages to agents for exerting an impact on their behavior [124]:

1. Bare facts
2. Logical inferences and analytical judgments based on a definite set of facts
3. Appellative appraisals like good–bad, moral–immoral, and ethical–depraved

New information employed by agents to make decisions can be divided into the following categories:

4. Hard information, which includes only actual data and facts
5. Soft information, which includes forecasts and appraisals

One would definitely see an analogy between items 1 and 4 as well as between items 2 and 5. This issue will be analyzed in the sequel; now we concentrate on item 3.

This item, in medias res, relates to the ethical aspect of information (the ethical aspect of choice). Lefebvre (and the other investigators who generalized his theory of ethical choice) made the only known attempt to provide a formal description of the ethical aspect. His approach proceeds from the assumption that a decision maker performs *first-kind reflexion*, that is, acts as an observer of his or her own behavior, thoughts, and feelings. The internal structure of an agent consists of several interconnected levels (in particular, a special level "being responsible for" the ethical aspect of choice). The final decision of an agent is determined by the mutual influence of an external environment and the states of these internal levels.

In game theory (and in this textbook as well), an agent is viewed as an individual (i.e., an "indivisible" person); he or she performs *second-kind reflexion*, which relates to decisions made by the opponents. Hence, we leave item 3 beyond our consideration and address items 1, 4 and 2, 5.

An awareness structure of agent i includes beliefs about:

1. The state of nature (θ_i);
2. Beliefs of opponents ($\theta_{i\sigma}$, $\sigma \in \Sigma_+$).

Reporting either θ_i or $\theta_{i\sigma}$ is an informational impact. In other words, a principal may report to an agent (or agents) information on the state of nature (i.e., the value of an uncertain parameter) and on the beliefs of the opponents.

Thus, we have the following types of informational impact:

(i) Informational regulation
(ii) Reflexive control

Roughly speaking, they correspond to items 1, 4.

Concerning items 2 and 5, we should underline that they correspond to the following type of informational impact:

(iii) Active forecast

It consists in reporting information on the future values of specific parameters that depend on the state of nature and actions of the agents.

5.4 Applied Models of Informational Control

This section contains several stylized examples of the models of informational control from the monograph [124]. The referenced book contains more than 30 examples of informational and reflexive control application to organizational, economic, social and other systems, and military applications.

The Scarcity Principle

The book of the same title by American psychologist R. Cialdini [43] deals with description and classification of stereotypes often followed by people in their behavior (when they make certain decisions). These stereotypes represent a kind of programs being launched under specific circumstances and determine human actions, including obviously irrational actions. In particular, six fundamental psychological principles that direct human behavior are identified: reciprocation, consistency, social proof, liking, authority, and scarcity. Let us discuss the last principle.

The idea of the *scarcity principle* is the following: Opportunities seem more valuable to us when they are less available. In particular, this is the case for scarce information, and exclusive information is more persuasive information. The following experiment is described in corroboration of these words; it was conducted by a successful businessman interested in psychology, the owner of a beef-importing company in the United States.

> The company's customers—buyers for supermarkets and other retail food outlets—were called on the phone as usual by a salesperson and asked for a purchase in one of three ways. One set of customers heard a standard sales presentation before being asked for their orders. Another set of customers heard the standard sales presentation plus information that the supply of imported beef was likely to be scarce in the upcoming months. A third group received the standard sales presentation and the information about a scarce supply of beef, too; however, they also learned that the scarce supply news was not generally available information—it had come, they were told, from certain exclusive contacts that the company had.
>
> ... Compared to the customers who got only the standard sales appeal, those who were also told about the future scarcity of beef bought more than twice as much...The customers who heard of the impending scarcity via "exclusive" information... purchased six times the amount that the customers who received only the standard sales pitch did. Apparently, the fact that the news about the scarcity information was itself scarce made it especially persuasive. [43, p. 214]

Not doubting the correctness of this conclusion, we endeavor to take a different view of the situation. Notably, we explain the actions of company's customers based on a game-theoretic model.

Thus, there exist n customers (called agents) of the company that make decisions regarding the amount of beef purchase. Suppose the number of the agents n is sufficiently large, while all agents are identical and compete in the Cournot framework. The price linearly depends on the supply; that is, the goal functions of the agents have the form

$$f_i(x_1,\ldots,x_n) = \left(Q - \sum_{j\in N} x_j\right) x_i - cx_i,$$

where $x_i \geq 0$, $i \in N = \{1, \ldots, n\}$, $c \geq 0$. These functions have the following interpretation: x_i means the sales volume of the agent during the period considered, $(Q - \sum_{j\in N} x_j)$ is the corresponding market price, and c denotes a wholesale purchase price. Then the first term of the goal function is a revenue (as the price is multiplied by the sales volume), and the second one expresses product purchase costs.

Next, we involve the first-order necessary optimality conditions to evaluate the following equilibrium actions of the agents (under the conditions of common knowledge):

$$x_i = \frac{Q - c}{n + 1}, \quad i \in N. \tag{5.16}$$

Recall all agents are identical according to the introduced assumption; therefore, they have the same equilibrium actions. This is the case in the absence of informational impact. Having received standard offers, the first group of agents purchase the product in the volume (5.16); the agents expect selling it during the given period.

Now, study the behavior of the second group (these agents are informed of the coming reduction in beef deliveries). It seems possible to assume that the agents consider this information as common knowledge. In this case, a rational action of the agents is purchasing a doubled volume of the product; the agents would aim to sell the product in the same equilibrium amount (5.16) during the next period (simultaneously, they would search for alternative suppliers).

Finally, consider the behavior of the third group. These agents have been informed of the coming reduction in beef deliveries (and that this information is available to a few agents only). Probably, it would be rational to adopt the following assumption for such agents. There are two types of agents, notably, uninformed and informed ones (outsiders and insiders, respectively). Evidently, the agents of the third group believe they are insiders. During the given period, uninformed agents will sell the product in the volume (5.16); being short of the product, they will not participate in the game in the next period. Thus, the number of players during the next period (in fact, coinciding with the number of insiders) goes down from n to a certain $k\,n$ ($k < 1$ indicates the percentage of insiders). Hence, during the next period the equilibrium action will be defined by

$$x_i' = \frac{Q - c}{kn + 1}. \tag{5.17}$$

Compare formulas (5.16) and (5.17) to observe the following fact. Under a sufficiently large n, we have

$$\frac{x_i'}{x_i} = \frac{n+1}{kn+1} \approx \frac{1}{k}.$$

This is why the agents of the third group purchase the product in the volume $(x_i + x_i')$, i.e., $(1/k + 1)$ times greater than the ones belonging to the first group. For instance, suppose that the percentage of outsiders (the agents of the third group) is equal to 20%. In other words, $k = 1/5$ and this is common knowledge). One obtains

$$x_i + x_i' = 6x_1.$$

Then it appears rational for the third group to purchase a volume that is six times higher than the first group.

Lump-Sum Payment

Consider an organizational system composed of a principal and n agents performing a joint activity.

The strategy of agent i lies in choosing an action $y_i \in X_i = \Re_+^1, i \in N$. On the other hand, the principal chooses an incentive scheme that determines a reward of every agent depending on the result of his or her joint activity. Imagine the following scheme of interaction among the agents. Attaining a necessary result requires that the sum of their actions is not smaller than a given threshold $\theta \in \Omega$. In this case, agent i yields a fixed reward σ_i, $i \in N$; if $\sum_{i \in N} y_i < \theta$, the reward vanishes for all agents.

Implementation of the action $y_i \geq 0$ makes it necessary that agent i incurs the costs $c_i(y, r_i)$, where $r_i > 0$ denotes his or her type (a parameter representing individual characteristics of the agent), $i \in N$.

Concerning the cost functions of the agents, let us assume that $c_i(y, r_i)$ is a continuous function, increasing with respect to y_i and decreasing with respect to r_i, and $\forall\ y_{-i} \in X_{-i} = \prod_{j \in N \setminus \{i\}} X_j$, $\forall\ r_i > 0$: $c_i(0, y_{-i}, r_i) = 0$, $i \in N$.

The stated model of interaction will be known as the *lump-sum payment*. Now, define the set of individual rational actions of the agents:

$$IR = \left\{ y \in X' = \prod_{i \in N} X_i \mid \forall i \in N, \quad \sigma_i \geq c_i(r_i) \right\}.$$

If the agents' costs are separable (i.e., the costs $c_i(y_i, r_i)$ of every agent depend on his or her actions only), we obtain $IR = \Pi_{i \in N}[0; y_i^+]$, where

$$y_i^+ = \max\{y_i \geq 0 \mid c_i(y_i, r_i) \leq \sigma_i\}, \quad i \in N.$$

Introduce the following notation:

$$Y(\theta) = \left\{ y \in X' \mid \sum_{i \in N} y_i = \theta \right\},$$

$$Y^*(\theta) = \operatorname{Arg} \min_{y \in Y(\theta)} \sum_{i \in N} c_i(y, r_i).$$

Next, we analyze different variants of agents' awareness about the parameter $\theta \in \Omega$. Obviously, even a slight complexity of the awareness structure could lead to substantial changes in the set of informational equilibria within the reflexive game.

Variant I

Suppose that the value $\theta \in \Omega$ is common knowledge. Then the equilibrium in the game of the agents is provided by a parametric Nash equilibrium belonging to the set

$$E_N(\theta) = IR \cap Y(\theta). \tag{5.18}$$

We also define the set of Pareto-efficient actions of the agents:

$$Par(\theta) = IR \cap Y^*(\theta). \tag{5.19}$$

Since $\forall\ \theta \in \Omega$: $Y^*(\theta) \subseteq Y(\theta)$, the expressions (5.18)-(5.19) imply that the set of Pareto-efficient actions represents a Nash equilibrium. However, the set of Nash equilibria may be wider; in particular, under $\theta \geq \max_{i \in N} y_i^+$ it always includes the vector of zero actions.

Assume that the cost functions of the agents are, in fact, the Cobb-Douglas functions: $c_i(y_i, r_i) = r_i\, \varphi(y_i / r_i)$, where $\varphi(\cdot)$ is a smooth strictly increasing function such that $\varphi(0) = 0$.

In this case, the point $y^*(\theta) = \{y_i^*(\theta)\}$, such that $y_i^*(\theta) = \theta r_i / \sum_{j \in N} r_j$, $i \in N$, is a unique Pareto-efficient point.

Let us evaluate $y_i^+ = r_i\, \varphi^{-1}(\sigma_i/r_i)$, $i \in N$. Then under the condition

$$\sigma_i \geq r_i\, \varphi\left(\theta / \sum_{j \in N} r_j\right), \quad i \in N, \tag{5.20}$$

the Pareto set is nonempty.

The sets of Nash equilibria in the game of $n = 2$ agents (with $\theta_2 > \theta_1$) are illustrated by Figure 5.10; note that the point (0; 0) is a Nash equilibrium in both cases.

Thus, we have studied an elementary variant of agents' awareness structure (corresponding to a situation when the value $\theta \in \Omega$ is common knowledge). Now, consider the next

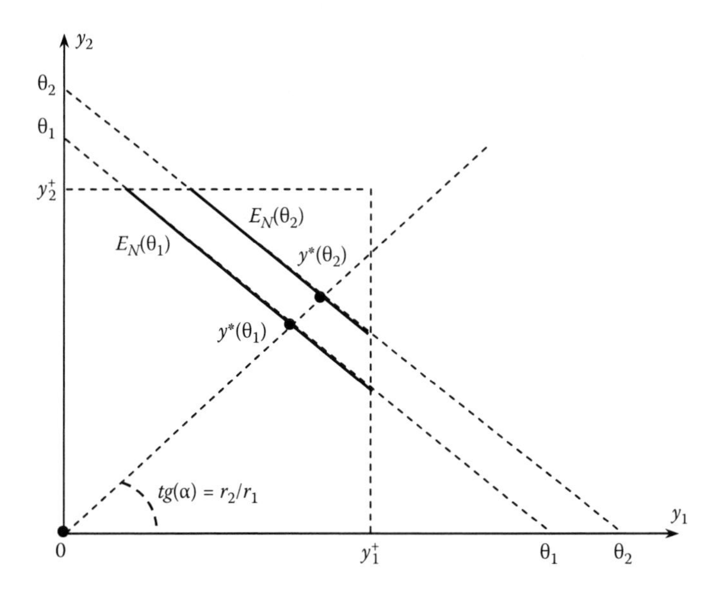

Figure 5.10 The parametric Nash equilibrium in agents' game.

variant with a greater level of complexity of agents' awareness; here common knowledge is represented by the individual beliefs $\{\theta_i\}$ of the agents about the value $\theta \in \Omega$.

Variant II

Suppose the beliefs of the agents about the uncertain parameter are pairwise different yet make up common knowledge. In other words, an asymmetric common knowledge takes place.

Without loss of generality, we renumber the agents so their beliefs form an increasing sequence: $\theta_1 < \ldots < \theta_n$. In this situation, the structure of feasible equilibria is defined in the following statement.

Assertion 5.4

Consider a lump-sum payment game, where $\theta_i \neq \theta_j$ for $i \neq j$. Depending on the existing relation among the parameters, the following $(n + 1)$ outcomes may be an equilibrium: $\{y^* \mid y_i^* = 0,\ i \in N\}$; $\{y^* \mid y_k^* = \theta_k,\ y_i^* = 0,\ i \in N,\ i \neq k\}$, $k \in N$. In practice, this means that either nobody works, or merely agent k does (and chooses the action θ_k).

Proof.

Let the action vector $y^* = (y_1^*, \ldots, y_n^*)$ be the equilibrium, i.e., $y_i^* \leq y_i^+$ for any $i \in N$. Assume there exists a quantity $k \in N$ such that $y_k^* > 0$. We demonstrate that in this case $\sum_{i \in N} y_i^* = \theta_k$.

Indeed, if $\sum_{i \in N} y_i^* < \theta_k$, then agent k does not expect any reward. Hence, he or she may increase the individual (subjectively expected) gain from a negative value to zero by choosing the zero action. On the other hand, with $\sum_{i \in N} y_i^* > \theta_k$, agent k reckons on a reward;

still, he or she may increase the individual gain by the action $\max\left\{0, \theta_k - \sum_{i\in N\setminus\{k\}} y_i^*\right\} < y_k^*$ instead of y_k^*. Therefore, if $\sum_{i\in N} y_i^* \neq \theta_k$, agent k may increase his or her gain; this fact contradicts the equilibrium character of the vector y^*.

We have shown that if $y_k^* > 0$ then $\sum_{i\in N} y_i^* = \theta_k$. But due to $\theta_i \neq \theta_j$, $i \neq j$, this equality holds for a single number $k \in N$. Consequently, $y_k^* > 0$ implies $y_i^* = 0$ for all $i \neq k$. And, clearly, $y_k^* > \theta_k$. •

Now, consider the issue regarding the relations among the parameters $\theta_i, y_i^+, i \in N$, that ensure every equilibrium stated in Assertion 5.4. The vector (0, …, 0) is an equilibrium when none of the agents may perform sufficient work independently (according to his or her view) for receiving a proper reward (alternatively, his or her action equals y_i^+ and the gain of agent i remains zero). Formally, the discussed condition is expressed by $y_i^+ \leq \theta_i$ for any i. The vector $\left\{y^* \mid y_k^* = \theta_k,\ y_i^* = 0, i \neq k\right\}$ is an equilibrium if $\theta_k \leq y_k^+$ and all agents with the numbers $i > k$ (actually, they believe no reward would be given) are not efficient enough for compensating the quantity $(\theta_i - \theta_k)$ themselves. Formally, we have $\theta_k + y_i^+ \leq \theta_i$ for any $i > k$.

Feasible equilibria in the game of two agents are presented in Figure 5.11. It should be emphasized that (in contrast to variant I) there exists an equilibrium-free domain.

To proceed, consider a general case when the beliefs of the agents may coincide: $\theta_1 \leq \ldots \leq \theta_n$. Similarly to variant I, this could lead to a whole domain of equilibria. For instance, suppose that $\theta_m = \theta_{m+1} = \ldots \theta_{m+p,}$ $\theta_i \neq \theta_m$ for $i \notin \{m, \ldots, m+p\}$. Under the conditions $\sum_{k-m}^{m+p} y_k^* \geq \theta_m$ and $\theta_m + y_l^+ \leq \theta_l, i > m$, the equilibrium is then given by any vector y^* such that $\sum_{k-m}^{m+p} y_k^* = \theta_m$, $y_k^* \leq y_k^+$, $k \in \{m, \ldots, m+p\}$; $y_i^* = 0, 1 \notin \{m, \ldots, m+p\}\}$.

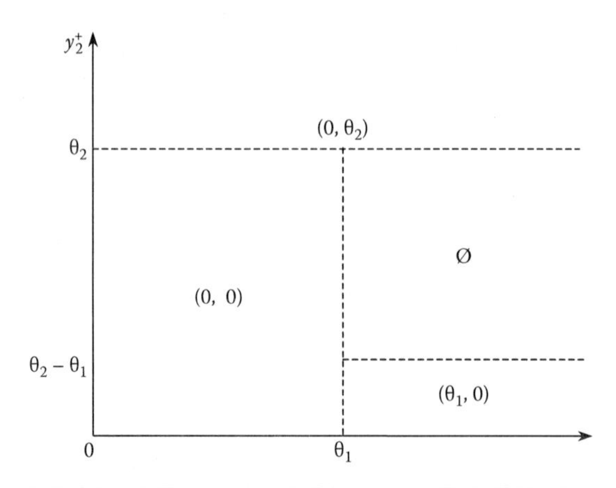

Figure 5.11 Equilibria in the game of two agents (the equilibrium-free domain is indicated by the symbol "∅").

The corresponding interpretation consists in the following. In an equilibrium, all work is performed by the agents with identical beliefs of the volume required for being rewarded.

Variant III

Let the awareness structure have depth 2 and each agent believe he or she participates in the game with an asymmetric common knowledge. In this case, the set of feasible equilibrium outcomes becomes maximal: $\Pi_{i \in N}[0; y_i^+]$. Moreover, the following statement takes place.

Assertion 5.5

Consider a lump-sum payment game. For any action vector $y^* \in \Pi_{i \in N}[0; y_i^+)$, there exists a certain awareness structure with depth 2 such that the vector y^* provides a unique equilibrium. Each agent then participates in the game with asymmetric common knowledge.

Proof.

For any $i \in N$, it suffices to select

$$\theta_i = \begin{cases} y_i^*, & y_i^* > 0 \\ y_i^+ + \varepsilon, & y_i^* = 0 \end{cases}$$

(here ε stands for an arbitrary positive number) and to choose any values $\theta_{ij} > \sum_{i \in N} y_i^+$, $j \in N \backslash \{i\}$. Then agent i expects zero actions from the opponents, while his or her own subjectively equilibrium action is given by y_i^*. •

Remark 1

Proving Assertion 5.5, we have constructed the equilibrium that is (objectively) Pareto-efficient if the sum $\sum_{i \in N} y_i^*$ equals the actual value of the uncertain parameter θ.

Remark 2

The action $y_i^* = y_i^+$ is an equilibrium provided that $\theta_i = y_i^+$. However, under this condition the action $y_i^* = 0$ forms an equilibrium, as well. In both cases, the subjectively expected gain of agent i makes zero.

Variant IV

Now, imagine the awareness structure of the game possesses depth 2 and the symmetrical common knowledge is at the lower level. In other words, every phantom agent believes that the uncertain parameter equals θ and this fact represents common knowledge.

It turns out that (even in this case) the set of equilibrium outcomes is maximal: $\Pi_{i \in N}[0; y_i^+]$. Furthermore, the following fact is true.

Assertion 5.6

Consider a lump-sum payment game with a symmetric common knowledge at the lower level. For any action vector $y^* \in \Pi_{i \in N}[0; y_i^+)$, there exists a certain awareness structure of depth 2 such that the vector y^* provides a unique equilibrium.

Proof.

Take any value $\theta > \sum_{i \in N} y_i^+$ and suppose this is common knowledge among the phantom agents. Then a unique equilibrium in the game of phantom agents is, in fact, the zero action chosen by every agent.

Next, for each number $i \in N$ we select

$$\theta_i = \begin{cases} y_i^*, & y_i^* > 0 \\ y_i^+ + \varepsilon, & y_i^* = 0 \end{cases}$$

with ε being an arbitrary positive number. Apparently, the best response of agent i to the zero actions of the opponents (he or she expects such actions) lies in choosing the action y_i^*. •

Remarks 1–2 (see variant III) expressis verbis remain in force for variant IV.

Therefore, we have studied the structure of informational equilibria in a lump-sum payment game under different variants of agents' awareness. The derived results corroborate the following (intuitively verisimilar) conclusion. Within a certain collective of employees, joint work is feasible (forms an equilibrium) if and only if there is common knowledge regarding the amount of works to be performed for being rewarded.

Now, let us address the issue of informational equilibrium stability. We will analyze variant II under an asymmetric common knowledge. Assume that, as the result of the game, common knowledge of the agents is the fact of paid or unpaid reward.

Clearly, the equilibrium (0, …, 0) is always stable; that is, nobody works, nobody expects a reward, and nobody receives a reward.

It has been demonstrated already that the equilibrium $\{y^* \mid y_k^* = \theta_k, y_i^* = 0, i \in N, i \neq k\}, k \in N$, where $\theta_1 < \ldots < \theta_n$, is feasible if $\theta_k \leq y_k^+, \theta_k + y_i^+ \leq \theta_i$ for any $i > k$. Then i-agents ($i \leq k$) expect a reward, and this is not the case for their colleagues with the numbers $i > k$. Hence, stability is guaranteed only by the condition $k = n$. Therefore, we obtain the following stability condition:

$$\theta_n \leq y_n^+. \tag{5.21}$$

Similarly, for $\theta_1 \leq \ldots \leq \theta_{m-1} < \theta_m = \ldots = \theta_n$ any set of the form

$$\left\{ y^* \mid \sum_{k=m}^{n} y_k^* = \theta_m,\ y_k^* \leq y_k^+,\ k \in \{m,\ldots,n\};\ y_i^* = 0,\ i \notin \{m,\ldots,m+p\} \right\}$$

appears stable.

According to Assertion 5.5, the principal may apply informational control for making the agents choose any set of actions $y^* \in \prod_{i \in N} [0; y_i^+)$ (e.g., this is achieved by forming a certain structure such that each agent subjectively plays a game with an asymmetric common knowledge). It turns out that a stable informational control exists that ensures the stated result. We demonstrate this in the case $y_i^* > 0$.

Suppose we have a given set $y^* \in \prod_{i \in N} (0; y_i^+)$, $\sum_{i \in N} y_i^* \geq \theta$. For each $i \in N$, set $\theta_i = y_i^*$; on the other hand, for each $j \in N \setminus \{i\}$ choose any θ_{ij} such that $\theta_{ij} < \theta_i$. Then i-agent subjectively satisfies the stability condition (5.21) and y_i^* represents his or her unique equilibrium action. The following should be noted:

1. The work will be completed, and the agents will receive rewards.
2. Receiving the rewards forms an expected outcome for all real and phantom agents.

The corresponding interpretation is the following. Every agent believes that exactly he or she has performed all work (and that this is common knowledge).

Corruption

Consider the following game-theoretic model of *corruption*. There are n agents (government officials), additional income of each official is proportional to the total bribe $x_i \geq 0$ taken by him, $i \in N = \{1, \ldots, n\}$. We will assume that a bribe offer is unbounded. Let every agent be characterized by the type $r_i > 0$, $i \in N$, which is known to him or her (and not to the remaining agents). The type may be interpreted as the agent's subjective perception of the penalty "strength."

Irrespective of the scale of corruption activity ($x_i \geq 0$), agent i may be penalized by the function $\chi_i(x, r_i)$, which depends on the actions $x = (x_1, x_2, \ldots, x_n) \in \Re^n_+$ of all agents and on the agent's type.

Consequently, the goal function of agent i is defined by

$$f_i(x, r_i) = x_i - \chi_i(x, r_i), \quad i \in N. \tag{5.22}$$

Suppose the penalty function has the form

$$\chi_i(x, r_i) = \varphi_i(x_i, Q_i(x_{-i}), r_i). \tag{5.23}$$

Formula (5.23) means that the penalty of agent i depends on his or her action and on the aggregated opponents' action profile $Q_i(x_{-i})$ (in the view of agent i, this is a "total level of corruption of the remaining officials").

Assume that the number of agents and the general form of the goal functions are common knowledge; moreover, each agent has a hierarchy of beliefs about the parameter $r = (r_1, r_2, \ldots, r_n) \in \Re^n_+$. Denote by r_{ij} the belief of agent i about the type of agent j, by r_{ijk} the belief of agent i about the beliefs of agent j about the type of agent k, and so on ($i, j, k \in N$).

In addition, assume that the agents observe the total level of corruption. Thus, informational equilibrium stability takes place under any beliefs about the types of real or phantom agents such that the corresponding informational equilibrium leads to the same value of the aggregate $Q_i(\cdot)$ for any $i \in N$. In this case, the goal functions (5.22)–(5.23) of the agents obviously satisfy the conditions of Assertion 5.3; hence, for any number of the agents and any awareness structure, all stable equilibria of the game are true. In other words, the following statement holds.

Assertion 5.7

Let the set of actions x_τ^*, $t \in \Sigma_+$, be a stable informational equilibrium in the game (5.22)-(5.23). Then this equilibrium is true.

Corollary

The level of corruption in a stable opponents' action profile does not depend on the mutual beliefs of corrupt officials about their types. Whether these beliefs are adequate or not appears not important.

It is then impossible to influence the level of corruption only by modifying the mutual beliefs. Therefore, any stable informational equilibrium results in the same level of corruption.

Suppose that

$$\varphi_i(x_i, Q_i(x_{-i}), r_1) = x_i(Q_i(x_{-i}) + x_i)/r_i,\ Q_i(x_{-i}) = \sum_{j \neq i} x_j,\ i \in N$$

and all types are identical: $r_1 = \ldots = r_n = r$. Evidently, equilibrium actions of the agents are $x_i = r/(n + 1)$, $i \in N$, while the total level of corruption constitutes

$$\sum_{i \in N} x_i = \frac{nr}{n+1}.$$

The latter quantity may be changed only by a direct impact on the agents' types.

Bipolar Choice

Consider a situation when the agents belonging to an infinitely large community choose between two alternatives. For generality, these alternatives are said to be positive and negative poles. It could be a candidacy for an election (voting for or against it), a certain product or service (buying it or not), or an ethical choice (acting in a good or a bad manner).

To solve the control problem for the whole community, let us suppose that the choice of a specific agent is not important (due to the infinite number of the agents). Actually, the share of the agents choosing the positive pole appears relevant. This statement could be reformulated as follows. The action of the "aggregated" agent is the probability x related to his or her choice of the positive pole.

Introduce the following assumptions:

1. There exist n different types of the agents.
2. The share of the agents having type i makes α_i, $0 \leq \alpha_i \leq 1$.
3. The action of an i-type agent is defined by the *response function on expectation:*

$$\pi(p), \pi : [0, 1] \rightarrow [0, 1].$$

Note that p means the probability (expected by the agents) of the event that an arbitrary agent from the community would choose the positive pole. In other words, if an agent expects the share of the agents choosing the positive pole equals p, then his or her action x_i is determined by

$$x_i = \pi_i (p).$$

4. Assumptions 1–3 are common knowledge among the agents.

Denote by $x_i \in [0, 1]$ the action of an i-type agent. Consequently, the share of such agents choosing the positive pole constitutes $p = \sum_{j-1}^{n} \alpha_j x_j$.

We define a *bipolar choice equilibrium* as a set of the actions x_i such that

$$x_i = \pi_i\left(\sum_{j=1}^{n} \alpha_j x_j\right), i = 1, \ldots, n. \tag{5.24}$$

Making a digression, let us underline that formulas (5.24) represent a possible way to describe the bipolar choice. Another approach is developed by Vladimir Lefebvre. He proceeds from that a decision maker performs the *first-kind reflexion*, that is, observes his or her behavior, ideas, and feelings. In other words, several interrelated levels exist within the agent; his or her final decision depends on the impact of an external environment and the states of the aforementioned levels. In the present textbook, we consider an agent as an individual, that is, an "indivisible person" who performs the *second-kind reflexion* (regarding the decisions made by his or her opponents).

We get down to the discussion of bipolar choice equilibrium. Note that formulas (5.24) specify a self-mapping of the unit hypercube $[0, 1]^n$:

$$(x_1, \ldots, x_n) \rightarrow \left(\pi_1\left(\sum_{j=1}^{n} \alpha_j x_j\right), \ldots, \pi_n\left(\sum_{j=1}^{n} \alpha_j x_j\right)\right). \tag{5.25}$$

The functions $\pi_i(\cdot)$ being continuous (this seems natural), the mapping (5.25) is also continuous. Then, according to the fixed point theorem, the system (5.24) admits (at least) a single solution.

Consider the following example. There are three types of agents ($n = 3$), and their actions are given by the functions

$$\pi_1(p) \equiv 1, \ \pi_2(p) = p, \ \pi_3(p) \equiv 0.$$

In practice, this means that first-type agents choose the positive pole (regardless of anything), while third-type ones prefer the negative pole. The agents of the second type hesitate in their choice, and their actions coincide with the expected action of the whole community.

In this case, the system (5.24) is reduced to

$$x_1 = 1,\ x_2 = \alpha_1\ x_1 + \alpha_2\ x_2 + \alpha_3\ x_3,\ x_3 = 0.$$

This yields

$$x_1 = 1,\ x_2 = \frac{\alpha_1}{1-\alpha_2},\ x_3 = 0$$

(provided that $0 < \alpha_i < 1$, $i = 1, 2, 3$). At the same time, we have

$$p = \alpha_1 x_1 + \alpha_2 x_2 + \alpha_3 x_3 = \alpha_1 + \alpha_2 \frac{\alpha_1}{1-\alpha_2}. \tag{5.26}$$

Now, imagine that a principal is able to influence the situation; he or she strives for increasing the probability of positive choice made by the whole community (in fact, this is the quantity p). To succeed, the principal may exert a certain impact on the agents belonging to the second or the third group (i.e., the first-type agents choose $x_1 = 1$ themselves). Suppose the principal may have an impact on the third group by moving a certain share y of its members to the second group. This procedure requires some resources (e.g., costs) in the amount of $C_2 y$. The principal may also influence the second group, changing the beliefs of its members about the parameter α_3 (irrespectively of the actual value). In particular, the impact consists in the following belief to be formed for the second group: "the share x of the members belonging to the third group has joined the second group." The corresponding costs to form such belief are $C_1 x$.

In other words, the principal modifies either an actual or a "phantom" (fictitious) share of the agents having the third type. Note that the total quantity of the resources available to the principal (a budget) makes C.

Principal's problem is distributing the resource C (i.e., choosing the shares x and y) to maximize the probability p. Formally, the optimization problem of the principal is posed as follows (see (5.26)):

$$p(x, y) = \alpha_1 + (\alpha_2 + y\alpha_3)\frac{\alpha_1}{1-(\alpha_2 + x\alpha_3)} \to \max \qquad (5.27)$$

under the constraints

$$C_1\, x + C_2\, y \le C, \quad 0 \le x \le 1,\, 0 \le y \le 1. \qquad (5.28)$$

Evidently, (5.27) is reduced to that of maximization for the function

$$\varphi(x, y) = \frac{\alpha_2 + y\alpha_3}{1-(\alpha_2 + x\alpha_3)}.$$

The latter increases with respect to both arguments, x and y; thus, the first constraint in (5.28) becomes an equality. Hence, the problem has been reduced to the maximum evaluation problem for the function

$$\psi(x) = \frac{\alpha_2 + \alpha_3(C - C_1x)/C_2}{1-(\alpha_2 + x\alpha_3)} = \frac{1}{C_1C_2}\frac{\alpha_2C_2/C_1 + \alpha_3C/C_1 - x\alpha_3}{1-\alpha_2 - x\alpha_3}.$$

Apparently, the function $\Psi(x)$ is monotonically increasing (decreasing or constant) if the quantity

$$\frac{\alpha_2\tilde{N}_2}{\tilde{N}_1} + \frac{\alpha_3\tilde{N}}{\tilde{N}_1} - (1-\alpha_2) \qquad (5.29)$$

is positive (negative or zero, respectively).

Denote $k_1 = C_1/C$ and $k_2 = C_2/C$. Then the positivity condition for (5.29) takes the form

$$\alpha_3 > k_1 - \alpha_2 (k_1 + k_2). \tag{5.30}$$

In the sequel, we assume that $C_1 > C$ and $C_2 > C$. This means that the principal is unable to "turn" all agents of the third type into the second-type ones. At the same time, an optimal choice of the principal is when all available resource is invested to increase actual or fictitious share of the second-type agents (under the condition (5.30)).

The relationship between the optimal choice of the principal and the parameters (α_2, α_3) is illustrated in Figure 5.12.

The shaded domain in Figure 5.12 corresponds to the true condition (5.30); that is, an optimal strategy of the principal lies in consuming all available resource to modify the beliefs:

$$x = \frac{C}{C_1}, \quad y = 0. \tag{5.31}$$

Solution (5.31) describes a situation when the share α_2 of the second-type agents is sufficiently large. Figure 5.12 makes

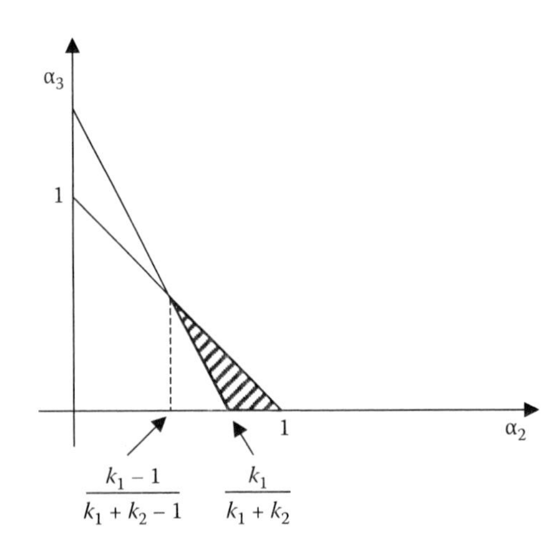

Figure 5.12 The optimal choice of the principal.

it clear that under $\alpha_2 > k_1/(k_1 + k_2)$ the solution (5.31) is always optimal. However, if

$$\frac{k_1 - 1}{k_1 + k_2 - 1} < \alpha_2 < \frac{k_1}{k_1 + k_2}, \tag{5.32}$$

the solution (5.31) is optimal for a sufficiently large α_3. In practice, this case possesses the following interpretation. For a certain range of the values α_2 (viz. the inequality (5.32) being true), it is optimal to impact the beliefs if they are too pessimistic; in particular, if α_3 is sufficiently large, the probability p of the negative pole choice is large, as well).

In conclusion, we emphasize that the elementary case of informational control under the conditions of bipolar choice has been studied here. Promising directions of future investigations include, first, further extension of the model (increasing the number of agents' types, making the awareness structure and the response function on expectation more complex) and, second, verifying the model using the results of the actions made by economic agents (customers) and political agents (voters).

Product Advertising

In this subsection we consider certain models of informational control implemented by the mass media. We involve *advertising* and the *hustings* as corresponding examples.

Suppose there is an agent representing the object of informational impact. The impact should form a required attitude of the agent to a specific object or subject.

In the case of advertising, the agent is a customer, and a certain product or a service acts as the object. The customer is required to purchase the product or service in question.

In the case of the hustings, a voter serves as the agent, and the subject is provided by a candidate. The voter is required to cast an affirmative vote for the given candidate.

Consider agent i. Combine the remaining agents to form a single agent (in the sequel, we will use the subscript j for him or her). Denote by $\theta \in \Omega$ an objective characteristic of the object (being unknown to all agents). For instance, the characteristics could be customer-related properties of the products or candidates' personal property.

Let $\theta_i \in \Omega$ be the beliefs of agent i about the object, $\theta_{ij} \in \Omega$ be his or her beliefs about the beliefs of agent j about the object, and so on.

For simplicity, let us make the following assumptions. First, the set of feasible actions of the agent consists of two actions, viz. $X_i = X_j = \{a; r\}$; the action a ("accept") means purchasing the product (service) or voting for the candidate, while the action r ("reject") corresponds to not purchasing the product (service) or voting for alternative candidates. Second, the set Ω is composed of two elements describing the object's properties: g ("good") and b ("bad"), that is, $\Omega = \{g; b\}$.

Now we study several models of the agent's behavior (according to the growing level of their complexity).

Model 0 (No Reflexion)

Suppose the behavior of the agent is described by a mapping $B_i(\cdot)$ of the set Ω (object's properties) into the set X_i (agent's actions); notably, we have B_i: $\Omega \to X_i$. Here is an example of such mapping: $B_i(g) = a$, $B_i(b) = r$. In other words, if the agent believes the product (candidate) is good, then he or she purchases the product (votes for this candidate); if not, he or she rejects the product (or the candidate).

Within the given model, informational control lies in formation of specific beliefs of the agent about the object, leading to the required choice. In the previous example, the agent purchases the product (votes for the required candidate) if the following beliefs have been a priori formed: $\theta_i = g$. Recall that the present textbook does not discuss any technologies of informational impact (i.e., the ways to form specific beliefs).

Model 1 (First-Rank Reflexion)

Suppose the behavior of the agent is described by a mapping $B_i(\cdot)$ of the sets $\Omega \ni \theta_i$ (object's properties) and $\Omega \ni \theta_{ij}$ (the beliefs of the agent about the beliefs of the remaining agents) into the set X_i of his or her actions, notably, $B_i: \Omega \times \Omega \to X_i$. The following mappings are possible examples:

$$B_i(g, g) = a,\ B_i(g, b) = a,\ B_i(b, g) = r;\ B_i(b, b) = r$$

and

$$B_i(g, g) = a,\ B_i(g, b) = r;\ B_i(b, g) = a,\ B_i(b, b) = r.$$

In the first case, the agent follows his or her personal opinion, while in the second case he or she acts according to the "public opinion" (the opinions of the remaining agents).

In fact, for the stated model informational impact is reflexive control. It serves to form agent's beliefs about the object and about the beliefs of the remaining agents, leading to the required choice. In the example considered, the agent purchases the product (votes for the required candidate) if the following beliefs have been a priori formed: $\theta_i = g$ with arbitrary θ_{ij} (in the first case) and $\theta_{ij} = g$ with arbitrary θ_i (in the second case).

Moreover, informational impact by the mass media not always aims to form θ_{ij} directly; in the majority of situations, the impact is exerted indirectly when the beliefs about behavior (chosen actions) of the remaining agents are formed for the agent in question. Consequently, the latter may use this data to estimate their actual beliefs. Examples of indirect formation of the beliefs θ_{ij} could be provided by famous advertising slogans like "Pepsi: The Choice of a New Generation" and "iPod: Everybody Touch." In addition, this could be addressing an opinion of competent people or revelation of information that (according to a public opinion survey) the majority of voters are going to support a given candidate.

Model 2 (Second-Rank Reflexion)

Suppose the behavior of the agent is described by a mapping $B_i(\cdot)$ of the sets $\Omega \ni \theta_i$ (object's properties), $\Omega \ni \theta_{ij}$ (the beliefs of the agent about the beliefs of the remaining agents) and $\Omega \ni \theta_{iji}$ (the beliefs of the agent about the beliefs of the remaining agents about his or her individual interests) into the set X_i of his or her actions, i.e., B_i: $\Omega \times \Omega \times \Omega \to X_i$. A corresponding example is the following:

$$\forall\, \theta \in \Omega\ B_i(\theta, \theta, g) = a,\ B_i(\theta, \theta, b) = r.$$

It demonstrates some properties being uncommon for Models 1 and 2. In this case, the agent acts according to his or her "social role" and makes the choice expected by the others.

For this model, informational impact is reflexive control; it consists in formation of agent's beliefs about the beliefs of the remaining agents about his or her individual beliefs (leading to the required choice). In the example above, the agent purchases the product (votes for the required candidate) if the following beliefs have been a priori formed: $\theta_{iji} = g$.

Note that informational impact not always aims to form θ_{iji} directly. In many situations the impact is exerted indirectly when the beliefs about expectations (actions expected from the agent) of the remaining agents are formed for the agent in question. The matter concerns the so-called social impact; numerous examples of this phenomenon are discussed in the textbooks on social psychology.

Indirect formation of the beliefs θ_{iji} could be illustrated by slogans like "Do you...Yahoo!?", "It is. Are you??", and "What the well-dressed man is wearing this year." Another example is revelation of information that, according to a public opinion survey, the majority of members in a social group (the agent belongs to or is associated with) would support a given candidate.

Therefore, we have analyzed elementary models of informational control by means of the mass media. The models have been formulated in terms of reflexive models of decision-making and awareness structures. All these models possess the maximum reflexion of 2. Probably, the exception is a rare situation when informational impact aims to form the whole informational structure (e.g., by thrusting a "common opinion" like "Vote with your heart!" and "This is our choice!").

It seems difficult to imagine real situations when informational impact aims at the components of the awareness structure that have a greater depth. Thus, a promising direction of further investigations lies in studying formal models of informational control (and the corresponding procedures) for the agents performing collective decision making under the conditions of interconnected awareness.

Now, assume there are two groups of agents. The first group tends to purchase the product regardless of advertising; the second one would rather not purchase it without advertising. Denote by $\theta \in [0; 1]$ the share of the agents belonging to the first group.

The agents in the second group (in fact, their share makes $(1 - \theta)$) are subjected to the impact of advertising; however, they do not realize this. We will describe the *social impact* in the following way. Suppose the agents in the second group choose the action a with probability $p(\theta)$ and the action r with probability $(1 - p(\theta))$. The relationship between the choice probability $p(\cdot)$ and the share of the agents tending to purchase the product reflects agents' reluctance to be a square peg in a round hole.

Suppose that the actual share θ of the agents belonging to the first group forms common knowledge. Then exactly θ agents will purchase the product; in fact, they observe that the product has been purchased by

$$x(\theta) = \theta + (1 - \theta)\, p(\theta) \tag{5.33}$$

agents (recall the agents do not realize the impact of advertising). Since $\forall \theta \in [0; 1]$: $\theta \leq x(\theta)$, an indirect social impact appears self-sustaining – "Look, it turns out that the product is purchased by more people than we have expected!"

Let us analyze asymmetric awareness. The agents of the first group choose their actions independently; hence, they could be treated as being adequately aware of the parameter θ and the beliefs of the agents in the second group.

Consider the model of informational regulation with the feature that a principal (performing an advertising campaign) forms the beliefs θ_2 of the agents in the second group about the value θ.

We digress from the subject for a while to discuss the properties of the function $p(\theta)$. Assume that $p(\cdot)$ is a nondecreasing function over [0; 1] such that $p(0) = \varepsilon$, $p(1) = 1 - \gamma$; here $\varepsilon \in [0; 1]$ and $\delta \in [0; 1]$ are constants such that $\varepsilon \leq 1 - \delta$). In practice, introducing the parameter ε means that some agents of the second group "make mistakes" and purchase the product (even if they think the remaining agents belong to the second group). In a certain sense, the constant δ characterizes agents' susceptibility to the impact. Notably, an agent in the second group has a chance to be independent; even if he or she thinks the remaining agents would purchase the product, he or she still has a chance to refuse. The special case $\varepsilon = 0$, $\delta = 1$ corresponds to independent agents in the second group that refuse to purchase the product.

The agents do not suspect the principle of strategic behavior. Hence, they expect to observe that θ_2 agents would purchase the product. Actually, it will be purchased by the following share of the agents:

$$x(\theta, \theta_2) = \theta + (1 - \theta)\, p(\theta_2). \tag{5.34}$$

Let principal's income be proportional to the share of the agents that have purchased the product. Moreover, suppose that the advertising costs $c(\theta, \theta_2)$ are a nondecreasing function of θ_2. Then the goal function of the principal (the income minus the costs) without advertising is defined by (5.33). With advertising being used, it is given by

$$\Phi(\theta, \theta_2) = x(\theta, \theta_2) - c(\theta, \theta_2). \tag{5.35}$$

Consequently, the efficiency of informational regulation may be specified as the difference between (5.35) and (5.33); the problem of informational regulation may be posed as

$$\Phi(\theta, \theta_2) - x(\theta) \to \max_{\theta_2}. \tag{5.36}$$

Now we discuss constraints of the problem (5.36). The first constraint is $\theta_2 \in [0;\ 1]$ (to be more precise, $\theta_2 \geq \theta$).

Consider an example. Set $p(\theta) = \sqrt{\theta}$, $c(\theta, \theta_2) = (\theta_2 - \theta)/2\sqrt{r}$, where $r > 0$ stands for a scaling factor. The problem (5.36) takes the form

$$(1 - \theta)(\sqrt{\theta_2} - \sqrt{\theta}) - (\theta_2 - \theta)/2\sqrt{r} \to \max_{\theta_2 \in [\theta;1]}. \tag{5.37}$$

Solution to the problem (5.37) is provided by $\theta_2(\theta) = \max\{\theta;\ r(1 - \theta)^2\}$; that is, under

$$\theta \geq \frac{(2r + 1) - \sqrt{4r + 1}}{2r}$$

informational regulation makes no sense for the principal (the advertising costs are not compensated, since a sufficient share of the agents purchase the product without advertising).

In addition to the constraint $\theta_2 \in [\theta;\ 1]$, let us require stability of informational regulation. Suppose that the share of agents purchasing the product is observable. Moreover, assume that the agents in the second group should observe the share of agents purchasing the product in a quantity not smaller than the value reported by the principal. That is, the stability condition could be rewritten as $x(\theta, \theta_2) \geq \theta_2$.

Use formula (5.34) to obtain

$$\theta + (1 - \theta)\, p(\theta_2) \geq \theta_2. \tag{5.38}$$

Hence, an optimal stable solution to the problem of informational regulation would be a solution to the maximization problem (5.36) under the constraint (5.38).

To conclude this subsection, we note that any informational regulation in the above example is stable in the sense of (5.38). Stability could be either viewed as a perfect coincidence between the expected results and the results observed by the agents (i.e., the constraint (5.38) becomes the equality). In this case, the only stable informational regulation is principal's message that all agents belong to the first group, that is, $\theta_2 = 1$. Such tricks are common for advertising.

Discussion

The methods of informational control described in this chapter are very subtle, sophisticated, and risky compared with the other methods. The matter is that an awareness (opinions, beliefs) is often difficult to model (in contrast to organizational structures or even an incentive scheme of a subject). Furthermore, the methods of informational control could be used as a kind of "superstructure" for any other methods. This is the case since applying control in an organizational system results in an equilibrium being dependent on the awareness of all participants of the system. On the other hand, exerting an impact on the awareness structure allows for equilibrium control; thus, performance efficiency of the whole system (in the view of a decision maker) could be improved.

Consider the model of decision making adopted in reflexive games. Actions of an agent are determined by nothing more than by his or her awareness regarding the state of nature and the beliefs of the opponents (other agents).

Therefore, how informational control of a principal impacts these beliefs is of crucial importance. In other words, the question consists in the following. How is informational structure of the game formed depending on a specific informational impact of the principal?

In this context, we claim that an exhaustive answer to the question seems impossible using only mathematical (in particular, game-theoretic) models. In the first place, the underlying reason is that acquisition of information by humans depends (to a large extent) on psychosocial factors. The secret of efficient informational control consists in addressing the unconscious, in using the ways to eliminate perception barriers and to overcome natural tolerance of humans to perception of new things.

It is obvious what difficulties arise when formalizing this process in the case of an intelligent rational decision maker. Such an agent is in the focus of investigations on game theory. All modern concepts of game solution are based (explicitly or implicitly) on the awareness structure at the starting moment of the game.* What happened before the game starts? How was a specific awareness formed? These issues are not actually treated. Probably, this is a boundary between a real man and an intelligent rational decision maker being modeled [124].

TASKS AND EXERCISES

5.1. A game involves three agents with the goal functions $f_i(\theta, x_1, x_2, x_3) = (\theta - x_1 - x_2 - x_3)\, x_i$, where $x_i \geq 0$, $i \in N = \{1, 2, 3\}$, and $\theta \in \Omega = \{1, 2\}$. Note this is a Cournot oligopoly with zero manufacturing costs.

* Multistep games are not the exception as well; during the play, awareness of the participants may be changed (including the case of informational exchange among the participants). Indeed, any multistep game has an initial moment and an initial awareness of the participants.

In practice, x_i represents a production output of agent i, θ describes the demand for the products. For compactness, we will call a pessimist any agent that considers the demand to be low ($\theta = 1$); accordingly, an optimist believes the demand is high ($\theta = 2$).

Draw the graph of the corresponding reflexive game and evaluate the informational equilibrium in the following cases:

5.1.1. Agents 1 and 2 are optimists and consider all agents as optimists with an identical awareness. Agent 3 is a pessimist and considers all agents as pessimists with an identical awareness.

5.1.2. Agents 1 and 2 are optimists and consider all agents as optimists with an identical awareness. Agent 3 represents a pessimist being adequately aware of agents 1 and 2.

5.1.3. Agent 1 believes common knowledge is that the demand would be low; agent 2 believes common knowledge is that the demand would be high. Agent 3 represents an optimist being adequately aware of agents 1 and 2.

5.1.4. Each agent is an optimist and considers the remaining agents to believe that the demand is low represents common knowledge.

5.2. A game involves two agents with the goal functions $f_i(\theta, x_1, x_2) = (\theta - x_1 - x_2)\, x_i$, where $x_i \geq 0$, $i = 1, 2$, and $\theta \geq 0$.

Give an example of informational structure with the following properties. Agent 1 is adequately aware of agent 2, the belief of agent 1 about the uncertain parameter is adequate. Agent 2 gains more in an equilibrium outcome than agent 1.

5.3. Give an example of outcome with three agents and an uncertain parameter θ such that actually $\theta = \theta_0$, while common knowledge among the agents is $\theta = \theta_1 \neq \theta_0$.

5.4. Give an example of outcome with three agents and an uncertain parameter θ such that actually $\theta = \theta_0$ and, if $\theta = \theta_0$ would be common knowledge among the agents, then all agents gain more in an equilibrium than in the initial outcome.

5.5. Analyze how the awareness impacts the actions and gains of the agents in Tasks 5.1-5.2.

5.6. Consider the framework of Exercise 5.2 and suppose that all agents are optimists. Agents 1 and 2 are aware of each other, as well as agents 2 and 3 are aware of each other. Agent 1 believes agent 3 considers all agents as pessimists with an identical awareness; according to agent 3, agent 1 also considers all agents as pessimists with an identical awareness. Draw the graph of the corresponding reflexive game. Evaluate an informational equilibrium.

5.7. Assume two countries (*A* and *B*) and a certain agent participate in a game. The agent is a high-ranking official of the country *A* and an intelligencer of the country *B* (the country *A* knows nothing about this). Draw the graph of the corresponding reflexive game.

5.8. Consider the framework of Exercise 5.7. Suppose the agent actually works for the country *B* and reports to the country *B* the data intentionally distorted by the country *B*. Draw the graph of the corresponding reflexive game.

5.9. [124] Two agents play the hide-and-seek game in several rooms with different illumination levels. Agent 1 hides himself in a room, while agent 2 has to choose the room to seek for him. Illumination levels of all rooms are common knowledge. The agents use the following strategies. Under the remaining equal conditions, the seeker prefers to seek in the rooms having a higher illumination level (finding the hider is easier). The hider understands that the seeker has little chance of success in a room with a lower illumination level. The growing rank of reflexion means that an agent understands that this fact is clear to his or her opponent and so on. What are the maximal rational ranks of reflexion for the agents?

5.10*. [124] A seller and a buyer have a hierarchy of mutual beliefs about the price of a product. They should negotiate the price to make the deal. Construct the corresponding model. Evaluate an informational equilibrium.

5.11*. [124] Consider the model of active expertise (see Section 4.5). An organizer of the expertise has an opportunity to form the beliefs of the experts about their opinions. What range of collective decisions could be implemented as an informational equilibrium of the reflexive game among the experts?

5.12*. [124, 127] Construct and study the model of team building; a team represents a set of agents such that their choices are determined by the hierarchy of their mutual beliefs about each other.

5.13*. Consider the model described in Exercise 5.12. Analyze the impact exerted by agents' awareness regarding the actions and activity results of team members on the speed of team building.

5.14*. Consider the model of resource allocation (see Section 4.3). Find what mutual beliefs of the agents about the required quantities of resources lead to a stable informational equilibrium. When is the equilibrium true? Demonstrate that stable inadequate beliefs exist only for the agents that are not "dictators".

5.15*. [124] Give definitions and illustrative examples for the following terms:

Attainability set

Reflexion

First-kind reflexion

- Second-kind reflexion
- Informational reflexion
- Strategic reflexion

Reflexive game

Awareness structure

Informational equilibrium

- Stable informational equilibrium
- Unstable informational equilibrium
- True informational equilibrium
- False informational equilibrium

Observation function

Informational regulation

Reflexive control

Active forecast

Axiom of self-awareness

Phantom agent

Complexity of awareness structure

Depth of awareness structure

Reflexion rank

Graph of reflexive game

Bipolar choice equilibrium

Chapter 6

Mechanisms of Organizational Structure Design

In the present chapter we investigate the mathematical models of organizational structure design. In Section 6.1 we illustrate the basic ideas, aspects, and trade-offs of organizational hierarchy design using a simple model of control hierarchy over a technological network. In Section 6.2 we survey the economical literature on optimization models of hierarchical organization. In Section 6.3 we introduce the universal framework of organizational hierarchy optimization and provide some general results on the shape of the optimal hierarchy. Section 6.4 describes the efficient methods for optimization of tree-shaped structures.

6.1 Problems of Organizational Hierarchy Design

How Should Organizations Be Studied?

At first sight, the question seems simple and even strange (as organizations are analyzed since ancients times). However, the detailed consideration shows that the answer is not trivial.

Perhaps the matter is that organizations represent the most complicated, diversified, volatile, and, consequently, least investigated form of life.

The existing variety of types, classes, and forms of organizations inflates with a growing rate, thus not allowing for development of a somewhat general concept or a theory. Actually, the most permanent and well-known types of organizations (e.g., family, ethnos, or country) have undergone significant changes during recent decades; the theories providing description to these organizations are often contrary.

Concerning organizations related to production activity, one should emphasize that changes in them are a direct consequence of their existence (more specifically, consequence of extended reproduction). Growing rapidly in recent decades, global network (virtual) organizations that form electronic communities and create electronic economy and electronic culture (i.e., a global electronic society) represent a symbol of global changes in the nature of organizations.

Formal models of a hierarchical organization have been developing since the early 1950s. First of all, they were caused by the demand for control in economic, social, and military organizations (which became more sophisticated). Second, the new scientific methodology—the *systems approach* or *systems analysis*—yielded new ways to analyze complex systems. Since that time, organizations have represented a field of application and the source of new problem statements for different mathematical methods (e.g., optimization, operations research, and game theory).

The computer revolution of that epoch provided the corresponding technological basis for the new mathematical methods, that is, mathematical modeling with numerical experiment as a research framework. Simulation of organizations' functioning became a task of numerical experiments. Two types of organization models have been designed to date: economic and engineering.

During the first half of the twentieth century, a continuous process of economic science formalization took place. It led to the development of a full-scale mathematical theory of market equilibrium. However, very soon it became obvious that the theory, first, was unable to explain many phenomena observed at the existing markets and, second, scarcely considered natural laws of internal organization in economic subjects (firms) [45]. Advances in economic science in the second half of the twentieth century made it possible to understand the relevance of informational aspects in functioning of economic systems. For instance, asymmetric information of agents and their limited capabilities of information processing and decision making were among these aspects. Neoclassic economic theory, inter alia, enabled elucidating the role and place of hierarchical organizations in the processes of manufacturing and welfare distribution.

Simultaneously with the development of *mathematical economics*, the first half of the twentieth century was remarkable for a rapid progress in *automatic control theory*. Advances in aircraft and spacecraft engineering (required design and operation of complex technical systems) generated an essential need in formal models to describe organization of their operation and development. Modeling of a complex technical system is impossible without its decomposition into simpler subsystems; it allows for analyzing the behavior of isolated subsystems and describing their interconnections. As the result of a multilevel decomposition, a complex object is represented as a certain hierarchy of embedded (simpler) components that determine the object's structure. The choice of the system structure exerts a major impact on performance parameters of the whole system. Thus, the number of investigations in the field of optimization methods for technical systems structure demonstrated stable growth. Successful application of the research results in the design and control problems for real technical systems engendered intention to use them for organizational and biological systems. This was done when new scientific directions—*cybernetics* and *systems theory*—were created.

Nowadays, economic and engineering approaches to organizational modeling are gradually growing closer, with the help of advances in information technology and computer science. As it turns out, operation of distributed computer systems (related to data processing) resembles activity of managers within organizations. To model organizational hierarchies, today many economists involve the terminology and results from engineering sciences (in particular, information science). Hence, it seems possible to speak about synthetic theories combining the advantages of the engineering and economic approaches.

Any economic system* includes a set of employees (*agents*) organized in a certain way. Owing to their organization, the employees act according to specific procedures and rules (*mechanisms*) to achieve the goal of the system.

Specialization of employees within an *organization* improves their efficiency in comparison with separate (unorganized) agents. Nevertheless, interaction of employees with different specialization should be coordinated in order to achieve the common goal. This is a fundamental problem of any organization, since coordination requires some efforts to plan mutual actions, to control the results, and to coordinate the goals of specific employees. Management functions in organizations are implemented by means of a *hierarchy*.†

On one hand, a hierarchy improves efficiency of employees' interaction (e.g., via planning and controlling the material flows, informational flows, and the other flows). On the other hand, implementation of management functions causes some costs. In modern economic systems the share of managers (performing only management functions) in the whole staff of

* This chapter uses the terms *organization* and *economic system* synonymously.

† Employees at upper levels of the hierarchy possess primary rights as against the ones at lower levels. This feature allows for achieving the goal of the system (even in case of conflicts among employees).

the company reaches 40% (see, e.g., [136]). Consequently, the key factor defining the efficiency of an economic system lies in the optimal management hierarchy.

Real organizations provide very limited opportunities for experiments with their management structure. Hence, mathematical models are required, first, to support efficient organizational hierarchy design and, second, to justify the need and the course of reorganization (under varying external conditions).

Classifying the Models of Hierarchical Structures

The approaches used to formulate and solve problems of organizational hierarchy design are rather diversified, not in the least due to complexity of the object described. A classification would assist in guiding through the whole variety of the models. Publications in the field provide several principles employed to systematize the models of organizational structure design. For instance, a series of classifications are based on formal characteristics of the models (e.g., the mathematical techniques used or classes of the considered structures).

A popular classification divides models of organizational structures design into four basic approaches. The first approach is based on constructing a decomposition graph for the goals and tasks of an organization.* The second one proceeds from that an organization should maximize a certain efficiency criterion (the *goal function* of the organization). This function being complex, the maximization problem has to be decomposed into subproblems; their solutions are provided by separate departments of the organization. Organizational structure design is then reduced to finding a feasible decomposition minimizing the efficiency losses. The third approach

* General systems theory states that an organizational structure, to a large extent, depends on the structure of this graph. The problem of organizational structure design is then reduced to the allocation problem, that is, distribution of subgoals over the departments and employees of the organization.

aims to construct a certain function specifying directly the relationship between operational efficiency of the organization and structural characteristics of the organizational hierarchy. Consequently, one should seek for a hierarchy that maximizes or minimizes the function considered. Finally, the fourth approach is connected with quantitative evaluation of the interconnections among the system elements and a hierarchical group by combining strongly connected elements within a single unit.

Another possible classification bases on formal characteristics of the models (e.g., the aim of the research, purposefulness of the whole system and its separate elements, uniformity of the elements, and the number of levels within an organizational structure). This detailed classification allows for dividing the whole set of models into numerous categories. Moreover, it is also possible to analyze the models (e.g., study the degree of similarity among the models in the sense of various attributes of classification).

Several well-known systems of classification are based not on formal but on practical characteristics of the models. In fact, the most typical attribute of classification is the type of tasks performed by *managers* (as elements of a management hierarchy*). For example, Radner [136] provides the following list of what managers usually do:

1. Observe the environment and results of past actions
2. Process and transmit information
3. Make decisions
4. Monitor the actions of other employees in a firm
5. Hire and fire
6. Train and teach
7. Plan

* Managerial tasks can be the basis of classification for the models of hierarchy design because most of them consider only a single type of managerial activity (the most important one in the authors' view).

8. Solve problems
9. Exhort, persuade, set goals and values

The following section illustrates the basic terms and notions related to organizational structures. The model of a management hierarchy controlling a technological network would serve as an example.

The Problem of Management Hierarchy over a Technological Network [110]

Let $N = \{w_1, \ldots, w_n\}$ be a set of *workers* able to interact. Denote by w_{env} the *external environment* interacting with the workers. Typical workers are denoted by $w, w', w'' \in N$.

Any function of the form

$$f : (N \cup \{w_{env}\}) \times (N \cup \{w_{env}\}) \rightarrow \Re_+^p \tag{6.1}$$

is called a *flow function*. In other words, for each pair of workers $w', w'' \in N$ the vector $f(w', w'')$ defines the *intensity of flows* between w' and w''. This vector includes p nonnegative components. Every component determines the intensity of a specific interaction between the workers; one would identify different types of interaction (e.g., a material flow and an information flow). Therefore, the technology of interaction between the workers specifies the flow function f or the weighted *technological network* (N, f). The technological network is *undirected*, that is, $\forall w', w'' \in N$: $f(w', w'') = f(w'', w')$.

We say the relation exists between two workers if and only if the flow between them makes zero. Suppose that the network has no loops; that is, for each worker w the equality holds: $f(w, w) = 0$.

Consider the following example. Set $N = \{w_1, w_2, w_3\}$ and $p = 1$; this means we have three workers and one-dimensional flows. Assume that the technological network consists of four relations: $f(w_{env}, w_1) = \lambda$, $f(w_1, w_2) = \lambda$, $f(w_2, w_3) = \lambda$,

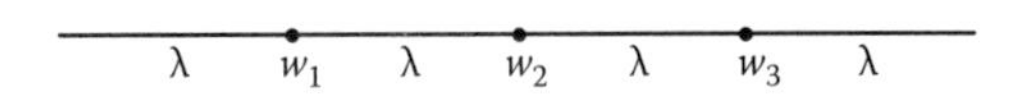

Figure 6.1 A symmetric production line.

$f(w_3, w_{env}) = \lambda$. Here λ stands for the flow intensity. This technological network is illustrated in Figure 6.1 and corresponds to a production line (a "business process"). Worker w_1 receives raw materials from a supplier, performs primary processing, and transfers the semimanufactured product to worker w_2. The latter performs the next technological operation and transfers the resulting semimanufactured product further, and so on. The final technological operation being finished, the last worker (in the present example, w_3) transfers the product to the customer.

A network with the workers $N = \{w_1, \ldots, w_n\}$ and the flows

$$f(w_{env}, w_1) = \lambda, \quad f(w_{i-1}, w_i) = \lambda$$

for any $2 \le i \le n$, $f(w_n, w_{env}) = \lambda$, is said to be a *symmetric production line.** The vector λ is referred to as *line intensity.*

Denote by M a finite set of *managers* controlling workers' interactions. Typical managers are referred to as $m, m', m'', m_1, m_2, \ldots \in M$.

Let $V = N \cup M$ be the set of all *employees* within an organization (workers and managers). Study the set of *subordination edges* $E \subseteq V \times M$. The edge $(v,m) \in E$ indicates that employee $v \in V$ is an *immediate subordinate* of manager $m \in M$, while m represents a *direct superior* of the employee v. Hence, an edge goes from an immediate subordinate to the immediate superior.

The employee $v \in V$ is a *subordinate* to the manager $m \in M$ (equivalently, the manager m is a *superior* to the employee v) if a chain of subordination edges is coming from v to m. We will also say that a superior *manages* a subordinate (equivalently, that a subordinate *is managed by* a superior). Let us state a rigorous definition of the hierarchy.

* The remaining flows are assumed zero.

Definition 6.1 [111]

The directed graph $H = (N \cup M,E)$ with the set of managers M and the set of subordination edges $E \subseteq (N \cup M) \times M$ is called a *hierarchy controlling the set of workers* N if the graph H has no cycles, any manager has subordinates, and there is a certain manager managing (immediately or not) all workers. The set of all hierarchies will be denoted by $\Omega(N)$.

The acyclic property means that no *circulus vitiosus* takes place, when each manager acts simultaneously as a superior and a subordinate to the other ones. Definition 6.1 eliminates numerous situations when managers possess no subordinates (this contradicts the manager's role, since he or she has to control certain employees).

The existence of a manager being a superior to all workers implies that any set of production workers enjoys a common boss; in other words, a hierarchy enables managing the interaction of all workers.

Figure 6.2 illustrates this definition. The workers are indicated by black spheres with Arabic numerals, while white spheres and Roman numerals are used for the managers.

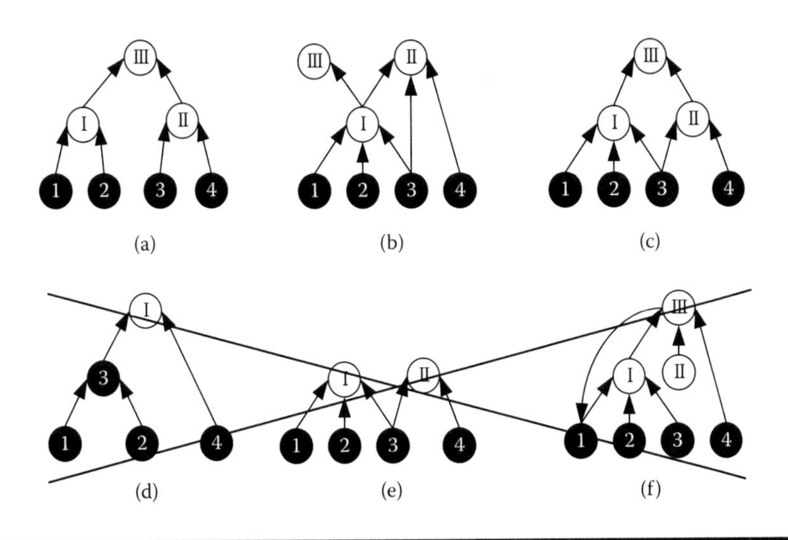

Figure 6.2 Illustrating the notion of hierarchy.

Graphs (a)–(c) demonstrate hierarchies controlling the set of workers $N = \{1, \ldots, 4\}$. The previous definition of hierarchy covers common management effects, such as interlevel interaction (a manager immediately controls both workers and other managers like manager II in hierarchy (b)) and multiple subordination (when an employee has more than one superior – manager I in hierarchy (b)) and worker 3 in hierarchy (c). Definition 6.1 admits of hierarchies with, first, several managers that have no superiors (managers II and III in hierarchy (b)) and, second, managers that have a single immediate subordinate (manager III in hierarchy (b)).

At the same time, graphs (d)–(f) are not hierarchies. Notably, worker 3 in graph (d) possesses subordinates, no top manager exists in graph (e) to control all workers, while manager II in graph (f) has no subordinates. (Moreover, this graph is cyclic: 1→I→III→1.)

In the sequel, we introduce some special classes of hierarchies. Moreover, we give an important definition of span of control.

Definition 6.2

A hierarchy is said to be a tree if it includes only a single manager m without superiors and the other employees have exactly one direct superior. The manager m is then called a *base* of the tree.

Figure 6.3a shows an example of a tree. On the contrary, the hierarchy in Figure 6.3b is not a tree (as it includes a manager having two direct superiors).

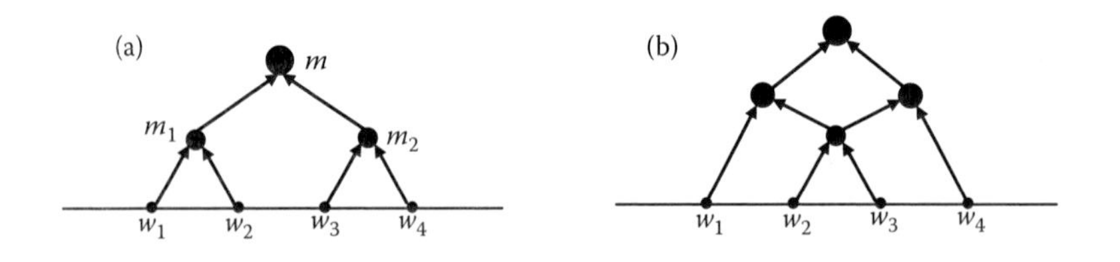

Figure 6.3 Examples of hierarchies over a production line.

Definition 6.3

A hierarchy is said to be an *r*-hierarchy if each manager has at most *r* direct superiors; here $r > 1$ means an integer number. If an *r*-hierarchy represents a tree, it will be referred to as an *r*-tree.

Many textbooks on management science utilize the term *span of control*; dating back to the works of Graicunas, it corresponds to the maximum number of immediate subordinates controlled by a single manager. The notion of *r*-hierarchy corresponds to the span of control being equal to *r*. According to Lemma 6.2, span of control does not exceed *n* for any tree. Note the maximum span of control among all trees is provided by a *two-level hierarchy* (a single manager directly controls all employees).

Definition 6.4

A hierarchy with a single manager directly controlling all workers is called a *two-level (fan) hierarchy*.

Definition 6.5 [111]

A hierarchy is said to be a *sequential hierarchy* if every manager directly controls (at least) one worker.

The area of the manager's responsibility (i.e., his or her role or duties) is a basic notion for organization structures. To formalize the manager's "role" within an organization, we state the notion of a *group being controlled by the manager*.

A *group* of workers $s \subseteq N$ is any nonempty subset of the set of workers.

Definition 6.1 implies that within any hierarchy *H* each manager has (at least) a single immediate subordinate. Starting from any manager *m*, we can go "down" the hierarchy to the subordinates of the manager *m*. As a result, it is possible to identify the set of workers being subordinated to the manager *m*. This set will be called a *subordinate group of workers* and denoted by $s_H(m) \subseteq N$. In addition, we will say that the manager *m controls the group of workers* $s_H(m)$. When mentioning the group $s_H(m)$, we omit the subscript if the hierarchy meant is clear from the context.

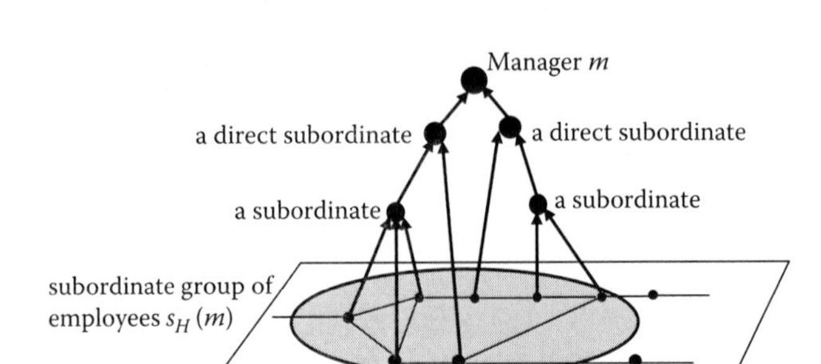

Figure 6.4 The manager and his or her subordinate group of workers.

Assume also that the worker $w \in N$ "controls" the elementary group $s_H(w) = \{w\}$ composed of this sole worker.

In Figure 6.4 the technological network is located in a horizontal plane. while the hierarchy grows upward. The section of the hierarchy illustrated above the plane is subordinated to the manager m. This section includes immediate subordinates of the manager m and the subordinates controlled by m indirectly. The subordinate group of workers $s_H(m)$ is outlined with the ellipse.

Let us formulate a simple lemma required for further development of the chapter.

Lemma 6.1 [110]

For any hierarchy H and any manager $m \in M$ the identity $s_H(m) = s_H(v_1) \cup \ldots \cup s_H(v_k)$ holds, where $v_1, \ldots, v_k$ are the immediate subordinates of the manager m. Any subordinate v of the manager m meets the inclusion $s_H(v) \subseteq s_H(m)$.

We explain this lemma using the following example. In Figure 6.3a the manager m has the managers m_1 and m_2 as immediate subordinates. Next, the group $s(m) = \{w_1, w_2, w_3, w_4\}$ is subordinated to the manager m. The managers m_1 and m_2 have the subordinate groups $s(m_1) = \{w_1, w_2\}$ and $s(m_2) = \{w_3, w_4\}$, respectively. Therefore, the group $s(m)$ is partitioned into two subgroups (denote them by $s(m_1)$ and $s(m_2)$) such that $\{w_1, w_2, w_3, w_4\} = \{w_1, w_2\} \cup \{w_3, w_4\}$. In the given example, the subgroups do not overlap. Generally (see Figure 6.3b), overlapping is possible.

Next, we state a useful lemma (the tree criterion for hierarchies).

Lemma 6.2 [110]

Consider a hierarchy H and suppose that only a single manager possesses no superiors. The hierarchy H represents a tree if and only if immediate subordinates of any manager control nonoverlapping groups of workers.

Thus, when only a single manager has no superiors, exactly a tree-type structure guarantees that immediate subordinates of any manager do not "double" duties of each other. (In other words, they do not control the same worker.)

In the model under consideration, the volume of work performed by a certain manager is determined, first, by the flows of technological network among workers of the group managed by him or her, and, second, by the flows between the controlled group and the rest of the organization. Any manager is assumed to perform duties of the two following types:

1. Managing the flows in a subordinate group, not controlled by the subordinate managers. For instance, the manager m controls the flow $f(w_2, w_3)$ in Figure 6.3a.
2. Participating in control of the flows between a subordinate group (on the one part) and the other employees and external environment (on the other part). In the previous expressions, this component is given in brackets; for example, in Figure 6.3a the manager m_1 participates in control of the flows $f(w_{env}, w_1)$ and $f(w_2, w_3)$.

Introduce a formal definition of managerial duties.

Definition 6.6 [110]

Within a hierarchy $H \in \Omega(N)$, the manager m performs duties of the following types:

1. Controlling the flows $f(w',w'')$ among the subordinate workers $w', w'' \in s_H(m)$ not controlled by any subordinate of the manager m. The sum of such flows will be referred to as the *internal flow* of the manager m and denoted by $F_H^{\text{int}}(m)$.
2. Participating in control of the flows $f(w',w'')$ between the subordinate worker $w' \in s_H(m)$ and the worker $w'' \in N \setminus s_H(m)$ which is not subordinated to the manager (and also between the worker $w' \in s_H(m)$ and the external environment $w'' = w_{env}$). The sum of such flows will be referred to as the *external flow* of the manager m and denoted by $F_H^{ext}(m)$.

Thus, a manager controls the internal flow and participates in control of the external one. The *flow of a manager* is the sum of his or her internal and external flows.

Definition 6.6 implies that the external flow of the manager m makes

$$F_H^{ext}(m) = \sum_{\substack{w' \in s_H(m), \\ w'' \in (N \setminus s_H(m)) \cup \{w_{env}\}}} f(w', w''). \tag{6.2}$$

The internal flow is evaluated using the following lemma.

Lemma 6.3 [110]

Suppose that $v_1, \ldots, v_k$ are all immediate subordinates of the manager m in a hierarchy H. Then the internal flow of the manager constitutes

$$F_H^{int}(m) = \sum_{\substack{\{w', w''\} \subseteq s_H(m), \\ \{w', w''\} \not\subseteq s_H(v_j) \text{ for all } 1 \le j \le k}} f(w', w''). \tag{6.3}$$

Therefore, to sum up the flows $f(w',w'')$ within the group $s_H(m)$, it suffices to check that they do not enter the groups controlled by immediate subordinates. Only in this case the flow would be not controlled by any subordinate of the manager (i.e., it would be included in the manager's internal flow).

For given values of N and f, we have that the internal and external flows of the manager m depend only on $s_H(v_1),\ldots,s_H(v_k)$. In other words, these flows depend on the groups of workers controlled by immediate subordinates of the manager m.

According to Definition 6.1, within any hierarchy H there exists a certain manager m who controls all workers (known as a top manager). Hence, *every flow within a technological network is controlled either by a top manager or by his or her subordinates.* This means that any hierarchy ensures control of all flows.

However, the number of managers and their workload vary in different hierarchies. Consequently, from the whole set of hierarchies $\Omega(N)$ one should choose a certain hierarchy being optimal in the sense of a specific criterion. We will consider the management cost as such a criterion (i.e., the total costs to keep all managers in the hierarchy). Within the framework of this basic model suppose that the total cost to keep a manager depends on the sum of the internal and external flows of this manager. Let us introduce a rigorous definition.

Definition 6.7

The costs of the manager $m \in M$ in the hierarchy $H \in \Omega(N)$ is the following function of the total controlled flow:

$$c(s_H(v_1),\ldots,s_H(v_k)) = \phi(F_H^{int}(m) + F_H^{ext}(m)). \qquad (6.4)$$

Here $v_1, \ldots, v_k$ indicate all immediate subordinates of the manager m, $s_H(v_1), \ldots, s_H(v_k)$ are the groups controlled by them, and $\phi : R_+^p \to R_+$ is a monotone function with respect to all its arguments, which maps the vector $F_H^{int}(m) + F_H^{ext}(m)$ into the costs of the manager.

The total costs of the whole hierarchy are, in fact, the sum of costs of the corresponding managers. An optimal hierarchy is the one minimizing the total costs. Again, we provide a formal definition.

Definition 6.8

We will call the costs of the hierarchy $H = (N \cup M, E) \in \Omega(N)$ the following sum of the costs incurred by all managers*:

$$c(H) = \sum_{m \in M} c(s_H(v_1), \ldots, s_H(v_k))$$
$$= \sum_{m \in M} \phi(F_H^{int}(m) + F_H^{ext}(m)), \tag{6.5}$$

where $v_1, \ldots, v_k$ represent all immediate subordinates of the manager m.

An optimal hierarchy is the hierarchy H^* with the minimum costs: $H^* \in \operatorname{Arg\,min}_{H \in \Omega} c(H)$.

We proceed from the assumption that (under known optimal hierarchy) one may hire managers that would perform the corresponding duties if their costs are compensated (e.g., by paying wages).† Of course, this requires complete information on the cost function. In what follows, we suppose that the manager's cost function $c(\cdot)$ is known. This function could be defined directly based on information regarding the costs of the managers. Moreover, it seems possible to consider some "typical" cost functions (e.g., below power-type functions are considered). Parameters of a specific function are chosen to render it as close to the actual costs of the managers as possible.‡

Evidently, even in elementary cases the set of different hierarchies is large, and finding the optimal hierarchy by exhaustive search requires overwhelming computational resources. Now we discuss some methods that ensure (under certain conditions) evaluation of the optimal hierarchy or narrow the set of hierarchies containing the optimal one.

---

* Note in formula (6.5) and in the sequel the same notation $c(\cdot)$ designates the cost function for the whole hierarchy and a specific manager.

† In addition to the manager's wage, the costs may also include some spendings intended to organize the manager's work (e.g., workplace, service personnel).

‡ The costs may be measured, such as in money equivalent of the efforts made (based on average wages of managers holding similar positions).

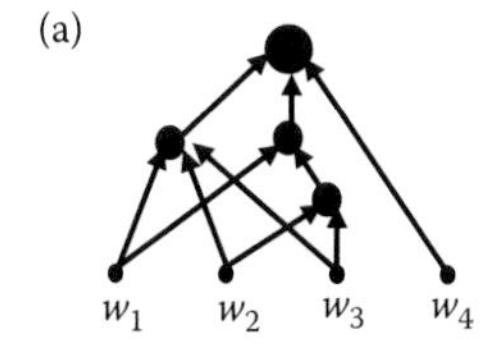

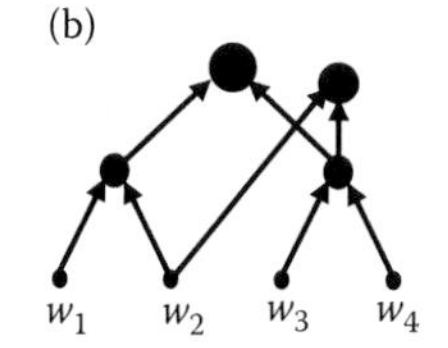

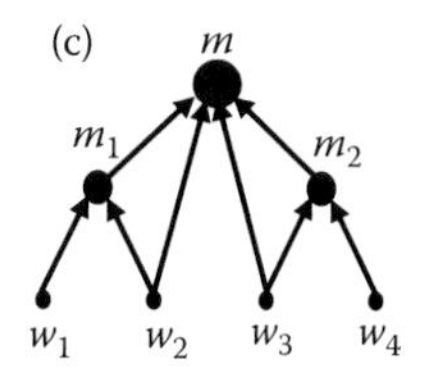

Figure 6.5 Hierarchies (a)–(c) violate properties (i)–(iii) of Assertion 6.1, respectively.

Assertion 6.1 [111]

For any hierarchy $H_1 \in \Omega(N)$ there exists a hierarchy $H_2 \in \Omega(N)$ possessing the same or smaller costs $(c(H_2) \leq c(H_1))$ and satisfying the following properties:

(i) All managers control different groups of workers.
(ii) Only a single manager possesses no superiors. The remaining managers and all workers are subordinated to this manager.
(iii) Immediate subordinates of the same manager do not control each other.

If H_1 represents an r-hierarchy, a tree, or an r-tree, then H_2 is an r-hierarchy, a tree, or an r-tree, respectively.

Proof.

This involves a consecutive reconstruction procedure for H_1 without costs increase. The reconstruction results in a certain hierarchy H_2 meeting conditions (i)–(iii). In the case of an r-hierarchy, a tree, and an r-tree, the procedure does not modify the type of hierarchy.*

Condition (i) means the absence of overlapping, when two managers control exactly the same group of subordinates. Figure 6.5a shows an example of such overlapping; two managers control the same group $\{w_1, w_2, w_3\}$. Note that one of these managers is easily eliminated (another manager is then made subordinated to all his or her direct superiors) without costs increase. In particular, condition (i) implies that *any manager possesses at least two immediate subordinates*

* We underline validity of Assertion 6.1 in general cases (see Section 6.3).

(otherwise, due to Lemma 6.1 he or she would control the same group as his or her single immediate subordinate).

According to condition (ii), there exists a single manager m who has no superiors. All workers $(s_{H_2}(m) = N)$ and the remaining managers within the hierarchy are subordinated to him or her. The manager m is said to be a *top manager.*

Condition (ii) conforms to a typical organization, where only the top manager has the right to make decisions being mandatory for all workers (e.g., negotiate a conflict situation between any workers). Figure 6.5b shows an example with two managers having no superiors; thus, condition (ii) is violated. Obviously, a "superfluous" manager could be eliminated without increasing the costs of hierarchy.

Condition (iii) may be interpreted in the following way. Suppose the manager m_1 is directly subordinate to the manager m. Then the latter does not directly control the subordinates of the manager m_1. This feature meets "regular" functioning of an organization, when managers control all workers by immediate subordinates (and not themselves). Figure 6.5c demonstrates a situation when the top manager m directly controls the workers w_2 and w_3 (despite they are controlled by the immediate subordinates m_1 and m_2 of the manager m). The edges (w_2, m) and (w_3, m) could be easily eliminated without costs increase of the hierarchy.

Assertion 6.1 implies there exists an optimal hierarchy satisfying conditions (i)–(iii).* This fact simplifies the problem of optimal hierarchy design; indeed, the hierarchies violating at least a single condition (i)–(iii) should not be considered.

In addition, Assertion 6.1 enables proving the following result. If there is an optimal r-hierarchy, an optimal tree, or an optimal r-tree, then there exists an optimal hierarchy of the corresponding type that meets conditions (i)–(iii). •

A simple condition ensures that the elementary two-level hierarchy is optimal.

* Consider H_1 as an optimal hierarchy. Assertion 6.1 states that the hierarchy H_2 meets properties (i)–(iii) and has the same or smaller costs. Hence, H_2 is an optimal hierarchy.

Assertion 6.2 [110]

Let the cost function $\phi(\cdot)$ be subadditive; that is, for all $x, y \in \Re_+^p$ the following inequality holds:

$$\phi(x + y) \leq \phi(x) + \phi(y).$$

Then the two-level hierarchy is optimal.

The subadditive property means that the costs $\phi(x + y)$ of a single manager controlling the total flow $x + y$ do not exceed the corresponding costs of two managers controlling the flows x and y independently. In this case, the elementary two-level hierarchy appears optimal (all flows are controlled by a single manager).

It follows from Assertion 6.2 that concavity of the cost function results in optimality of the two-level hierarchy if all flows within a technological network have the identical type (i.e., the flow vector includes a single component).

Two-level hierarchies (also referred to as *elementary structures*) are common in small organizations [109]. In a growing organization, a single top manager is very busy; therefore, the manager is compelled to hire assistants, that is, to change to a multilevel hierarchy.

Tree-shaped hierarchies are widespread in real organizations, but, in fact, they are not essentially optimal. The following example demonstrates that *sometimes the optimal hierarchy is not a tree.*

Example 6.1 (Costs Reduction under Double Subordination in an Asymmetric Line)

Assume an asymmetric production line has four workers and the following flows: $f(w_{env}, w_1) = 3$, $f(w_1, w_2) = 1$, $f(w_2, w_3) = 5$, $f(w_3, w_4) = 1$, $f(w_4, w_{env}) = 3$. Consider the cost function of a manager in the form $\phi(x) = x^3$ (here x is the manager's flow) (see (6.4)). The optimal hierarchy for this line is shown in Figure 6.7; denote it by H. The manager m_1 possesses two direct superiors; in other words, the hierarchy includes double subordination.

Let us evaluate the flows for every manager:

m_1: $c(\{w_2\}, \{w_3\}) = \phi[F_H^{int}(m_1) + (F_H^{ext}(m_1))] = [5 + (1 + 1)]^3 = 343;$

m_2: $c(\{w_1\}, \{w_2, w_3\}) = \phi[F_H^{int}(m_2) + (F_H^{ext}(m_2))] = [1 + (3 + 1)]^3 = 125;$

m_3: $c(\{w_4\}, \{w_2, w_3\}) = \phi[F_H^{int}(m_3) + (F_H^{ext}(m_3))] = [1 + (1 + 3)]^3 = 125;$

m_4: $(\{w_1, w_2, w_3\}, \{w_2, w_3, w_4\}) = \phi[F_H^{int}(m_4) + (F_H^{ext}(m_4))] = [0 + (3 + 3)]^3 = 216.$

Thus, the total costs of the hierarchy constitute

$$\begin{aligned} c(H) &= c(\{w_2\}, \{w_3\}) + c(\{w_1\}, \{w_2, w_3\}) + c(\{w_4\}, \{w_2, w_3\}) \\ &\quad + c(\{w_1, w_2, w_3\}, \{w_2, w_3, w_4\}) \\ &= 343 + 125 + 125 + 216 = 809. \end{aligned}$$

Let us prove that the computed costs deliver the minimum. Suppose H^* is the optimal hierarchy meeting conditions (i)–(iii) of Assertion 6.1. Note H^* includes (at least) a single lower-level manager m, which has no subordinate managers.

If the manager m controls three (or more) workers, then his or her flow is not smaller than 10. This means the manager's costs are not smaller than 1,000 (i.e., exceed the cost of the hierarchy H, $c(H) = 809$). Consequently, the manager m controls exactly two workers.

If the manager m controls two workers that are not adjacent in the production line (e.g., w_1 and w_3), then $F_{H^*}^{\text{int}}(m) = 0$. In other words, this manager controls no internal flow (merely participating in control of the external flows). Hence, the manager m could be eliminated, while the workers belonging to the group $s_{H^*}(m)$ could be made subordinate to the direct superiors of the manager m. Moreover, their costs would remain the same, and this fact contradicts optimality of H^*. Thus, only two adjacent workers in the production line could be subordinates of the manager m.

Imagine the workers w_1 and w_2 (alternatively, w_3 and w_4) are subordinated to the manager m; in this case, the costs of the latter constitute $9^3 = 729$. In addition, the top manager (at least) participates in control of the external flows and his or her costs are not smaller than $6^3 = 216$. Notably, we then have $c(H^*) > 729 + 216 = 945$ and again arrive at the contradiction to optimality of H^*. Therefore, H^* includes exactly

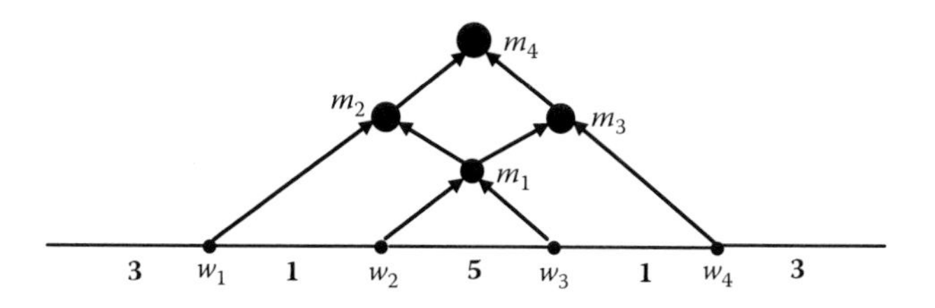

Figure 6.6 An example of optimal hierarchy, which controls an asymmetric production line.

one manager (m) at the lower level, controlling the workers w_2 and w_3 (i.e., the maximal flow $f(w_2, w_3) = 5$).

The example provides an illustration of the following general rule: the *flows with maximum intensity must be controlled at the lower levels of a hierarchy*. This rule is established in many publications on management science (see, e.g., [109]) as a result of intensive research in real organizations.

The previous example studies an extreme case when a special lower-level manager is allocated to control the maximum flow.

Since m is the only lower-level manager, he or she is subordinated to the other managers of the hierarchy.* Then the workers w_2 and w_3 are directly subordinated to the manager m. (If not, condition (iii) of Assertion 6.1 is violated.) In other words, the manager m being assigned, the optimal hierarchy H^* is built over three workers: w_1, m, w_4. In addition to the hierarchy H (see Figure 6.6), we obtain three alternative hierarchies meeting conditions (i)–(iii) of Assertion 6.1. They are shown in Figure 6.7.

It is easy to evaluate $c(H_1) = c(H_3) = 811$, $c(H_2) = 855$. Due to $c(H) = 809$, all hierarchies in Figure 6.7 are not optimal; hence, $H = H^*$ appears to be the unique optimal hierarchy.† •

Therefore, the current example demonstrates that, for a general technological network, *no optimal hierarchy exists*

* Any manager $m' \neq m$ has a manager m'' as an immediate subordinate (otherwise, m' represents a lower-level manager: $m' = m$). If $m'' \neq m$, this reasoning is repeated; thus, we arrive at the manager m and show his or her subordination to the manager m'.

† We mean the hierarchies satisfying conditions (i)–(iii) of Assertion 6.1.

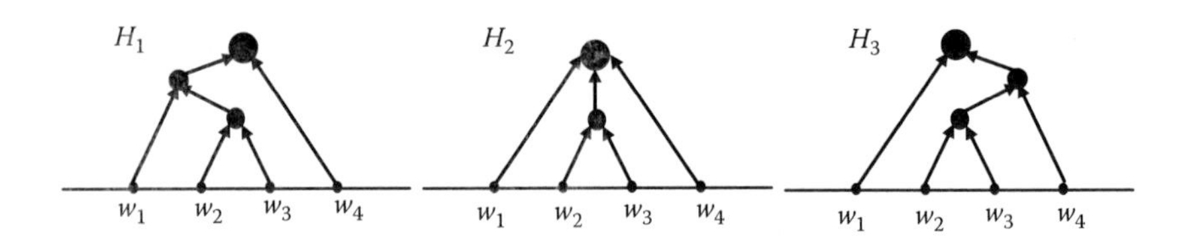

Figure 6.7 Suboptimal hierarchies over an asymmetric production line.

among trees. In what follows, we prove optimality of a tree-shaped structure for a symmetric production line. Also, a sufficient condition of tree-shaped structure optimality is derived in the following example within a broader framework. The problem of optimal tree search appears simpler, although absence of the general algorithm of polynomial complexity is proved [151].

Example 6.2 Management Expenses Decrease with Organization Expansion

Consider an asymmetric production line composed of four workers with the flows $f(w_{env}, w_1) = 1, f(w_1, w_2) = 5, f(w_2, w_3) = 1, f(w_3, w_4) = 5, f(w_4, w_{env}) = 1$ and the manager's cost function $\phi(x) = x^2$ (x is the manager's flow). First, assume that only the workers w_2 and w_3 belong to the organization; that is, consider the technological network $N = \{w_2, w_3\}$. Then there exists the sole hierarchy such that the conditions of Assertion 6.1 hold (see Figure 6.8a).

Now, suppose the organization is extended by adding two workers, w_1 and w_4, in the process of vertical integration. For instance, consider a large-scale wholesale company purchasing a raw materials supplier (the "worker" w_1) and

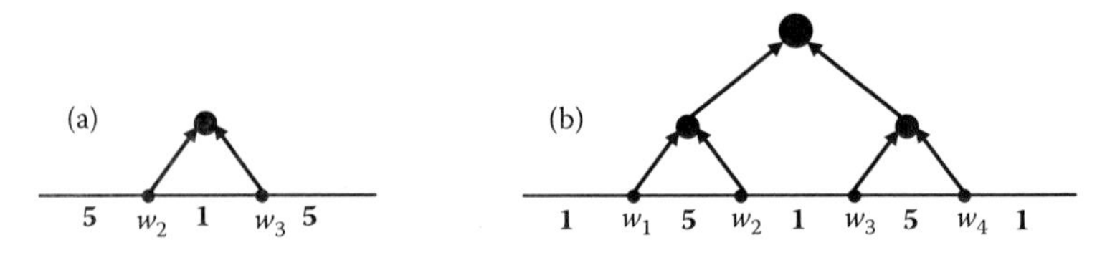

Figure 6.8 Organization expansion resulting in decrease of management expenses.

a retail store network (the "worker" w_4), striving to control the whole chain from the upstream to the downstream. The large flow $f(w_1, w_2) = 5$ is due to, for example, the information flow connected with the interaction problems between the company and the raw materials supplier (let's say, due to the high level of defects). Similarly, the large flow $f(w_3, w_4) = 5$ can be connected with interaction problems between the company and the retail store network (e.g., due to the high volume of product returns).

The extended organization controls the technological network $N = \{w_1, w_2, w_3, w_4\}$. It is possible to modify the control hierarchy as shown in Figure 6.8b (hire two lower-level managers, making them responsible for controlling the large flows). Compare the costs of the hierarchies:

$$\text{(a) } (5 + 1 + 5)^2 = 121,$$

$$\text{(b) } (1 + 5 + 1)^2 + (1 + 5 + 1)^2 + (1 + 1 + 1)^2 = 49 + 49 + 9 = 107.$$

Hence, the management expenses could be reduced by extending the technological network (adding new workers as a part of an external environment). This is a possible reason to purchase a new subsidiary business, which is unprofitable per se, yet helps reduce the management expenses of the primary business. Such a merger is a very common scenario in business. For instance, in the 1990s many Russian food industry enterprises were transformed into agroindustrial complexes with vertical integration by purchasing regional farms; the latter were not profitable yet provided regular supplies of components of guaranteed quality.•

Example 6.3 Multicomponent Flows

Assertion 6.2 implies that a two-level hierarchy is optimal under a concave cost function and single-component flows. Generally, this is not true for multidimensional flows. Consider a two-component flow ($p = 2$). The first component corresponds to a material flow, while the second one describes a flow of information. Suppose that $N = \{w_1, w_2, w_3, w_4\}$ and the technological network is as in Figure 6.9.

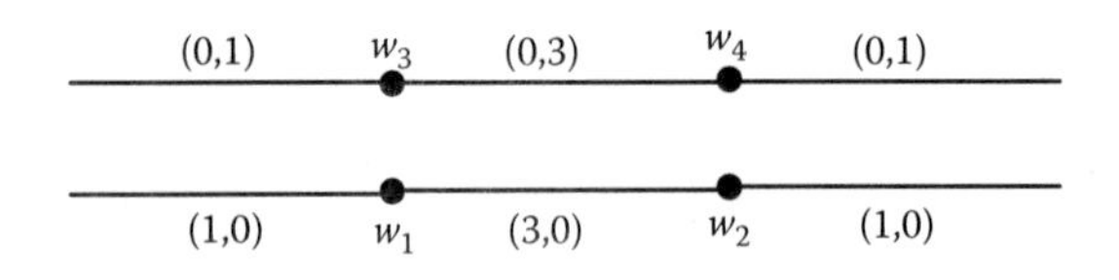

Figure 6.9 An example of a technological network with two-component flows.

The worker w_1 receives raw materials, performs a technological operation, and passes the semifinished product to the worker w_2; the latter assembles the finished product and ships it to a customer. The first component of the flow may, for example, correspond to the number of types of raw materials being passed. The worker w_1 receives raw materials of a single type and produces three types of parts. The worker w_2 receives these parts, assembles them into a single finished product, and ships to the customer. Hence, the internal flow $f(w_1, w_2)$ is larger than the external flows $f(w_{env}, w_1)$ and $f(w_2, w_{env})$. The second component of the flow stands for information and documents. The worker w_4 negotiates possible deals with the customers, prepares and concludes contracts of supply, and controls the corresponding payments and shipments of the products.

Information on necessary production volume is transferred to the worker w_3 by the worker w_4. Based on the acquired data, the worker w_3 places an order for raw materials, controls incoming materials, and performs different calculations. In addition, the worker w_3 may transfer to the worker w_4 information to evaluate the price and delivery period of the order. The internal information flow $f(w_3, w_4)$ could be larger than the external flows $f(w_{env}, w_3)$ and $f(w_4, w_{env})$, for example, due to numerous internal documents.

Assume that the manager's cost function has the form

$$\phi(x, y) = \sqrt{x} + \sqrt{y} + \sqrt{xy},$$

where (x, y) stands for the flow vector of the manager. This function is concave (yet, not subadditive), representing a situation when the managers are not very busy; thus, increasing the controlled flow reduces the costs per unit flow. The term $\sqrt{xy}$ reflects losses due to universalization of managers.

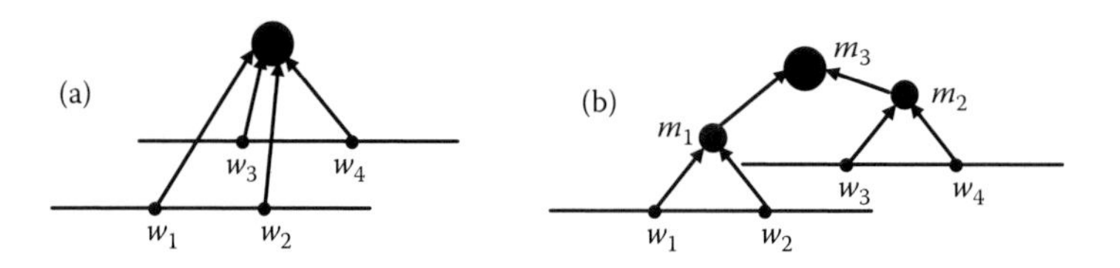

Figure 6.10 (a) A nonoptimal two-level hierarchy and (b) a hierarchy with specialized managers m_1 and m_2.

In particular, it equals zero if a manager controls a certain single-component flow (e.g., a production process or a document flow). In this case, the manager becomes a specialist in the corresponding area and controls the flow at minimum costs. If the manager has to control both types of the flow, the costs increase as the manager becomes less specialized.

Figure 6.10a illustrates the two-level hierarchy H_1. Here the flow of the sole manager makes (5, 5); that is, the cost of the hierarchy is $c(H_1) = \phi(5,5) = 2\sqrt{5} + 5$.

Analyze the following hierarchy H_2 with three managers (see Figure 6.10b). The manager m_1 controls only the production process, that is, the workers w_1 and w_2; his or her flow equals (5, 0). The costs of the manager m_1 constitute $\phi(5, 0) = \sqrt{5}$. Similarly, the manager m_2 controls only document flow, viz. the workers w_3 and w_4; his or her flow and costs are (0, 5) and $\phi(0, 5) = \sqrt{5}$, respectively. The top manager m_3 has the subordinate managers m_1 and m_2. The manager m_3 avoids going into the heart of the flows within the technological network. Instead, he or she merely participates in controlling the flows between the technological network and an external environment (interaction with customers and suppliers). The costs of the manager m_3 are given by $\phi(2, 2) = 2\sqrt{2} + 2$. Hence, $c(H_2) = 2\sqrt{5} + 2\sqrt{2} + 2$.

Finally, we obtain $c(H_2) < c(H_1)$. This means that *under multicomponent flows the costs of a hierarchy could be reduced by appointing several specialized managers even in the case of a concave cost function.*•

In some cases, the described model of technological flow control makes it possible to study analytically changes in a hierarchy that controls a relatively simple technological network. In particular, the optimality conditions for divisional structures, functional structure, and matrix structures have been investigated in [110]. Let us proceed to the systematic overview of the mathematical models of organizational structure design.

6.2 Models of Organizational Structure Design

In this section we survey mathematical models of organizational structure design. Following [63], the reviewed approaches are divided into *lines of research*, that is, the groups of related publications; the authors either develop the same model or discuss and criticize it. An obvious advantage of such an approach lies in historical authenticity since it enables tracing back the progress of approaches used to study the problems of organizational hierarchy design. Yet a drawback is that the survey becomes more eclectic. The following lines of research have been identified in the literature:

1. Multilevel symmetrical hierarchies
2. Hierarchies of knowledge
3. Data processing hierarchies
4. Hierarchies and the theory of teams
5. Decision-making hierarchies
6. Hierarchies and contract theory
7. A general model of optimal hierarchy search

Let us consider them in detail.

Multilevel Symmetric Hierarchies

This line of research is based on the model of the organizational structure as a sequence of hierarchical *levels of control*. In such hierarchies the length of the chain of command between any *worker* (located at the lower level of a hierarchy) and a *top manager* (located at the upper level) is the same; therefore, such hierarchies are said to be *symmetric*.

The problems studied within this line of research originate from the discussion in the economic literature during the first half of the twentieth century on the factors limiting growth of a firm [45]. The discussion resulted in the idea that the major factor was the limited individual ability of a firm owner to coordinate and control workers' activity. Thus, the corresponding authority required delegation to *middle-level managers.* The losses incurred by operation of the managerial hierarchy (not only the expenses to keep the managers but also the reduced efficiency due to the *loss of control*) are exactly the factors that could provide benefits to a large-sized firm (concentration of technologies and capital, leveled risks, and other factors of scale). However, are these losses as significant as to make infinite growth of a firm unprofitable? Answering this question required development of formal models of organizational hierarchies.

In the early paper by M. Beckmann [12] managerial structure of a firm is described by a sequence of hierarchical levels numbered top-down, starting from zero. There are L_i managers at level i, each obtaining the reward w_i for a fulfilled work. The ratio L_{i+1}/L_i (i.e., the numbers of managers at two adjacent levels) defines *span of control.* In fact, this rate represents the average number of *immediate subordinates* for each manager belonging to the level i.

In his well-known paper [155] the Nobel prize winner O. Williamson makes an inference that the control efficiency inevitably decreases for a growing firm. The matter is that a top manager acquires less information on the "old part" of the firm being extended (he or she should find time to get acquainted with information on the "new part"). As the result, the orders of the top manager become less detailed. Analyzing the model of firm's profit maximization (the difference between the income and costs), Williamson derives the approximate formula for the optimal number of levels within a hierarchy (the optimal size of a firm) resulting in the maximum profit.

G. Calvo and S. Wellisz [37] adhere to the opinion that the manager's main function within a hierarchy is to *monitor* the work efforts of his or her immediate subordinates. Thereby, loss of control (leading to a limited size of a firm) may take place or not, depending on specific features of the monitoring procedure employed. In this model the managers may introduce linear penalties for shirking of his or her immediate subordinates, while efficiency of monitoring depends both on the monitoring effort and span of control of the manager.

M. Keren and D. Levhari [81] analyze a model of a hierarchical firm where planning time (consequently, decision time) is defined by the total decision time at different levels of the hierarchy; hence, it directly influences the production output. Under rather natural assumptions regarding parameters of the model, it is demonstrated that the average *costs per unit product* increase as the firm grows. This means the growth limit actually exists.

To evaluate the optimal hierarchy, Keren and Levhari utilize the framework of optimal control theory in a quite elegant way. A detailed description of applying this framework to optimal hierarchy problems could be found in the later work [135] by Y. Qian. Qian's model integrates some features of the aforementioned models.

In a series of publications dedicated to the models of symmetric multilevel hierarchies, the paper by S. Rosen [138] seems to stand by itself. Instead of a separate firm, this work studies the whole *market*, which includes firms and managers.

Here are the conclusions for the research line studying the models of symmetric multilevel hierarchies:

1. Whatever the functions performed by managers are, the type of optimal hierarchy, the corresponding costs and operation efficiency of an organization essentially depend on the control mechanisms employed (e.g., planning, incentive, and controlling).
2. Under a rationally organized hierarchy of control, infinite growth of a firm is possible (to be more precise, the maximum size of a firm is limited by other factors not directly connected with organization maintenance costs, e.g., market demand).

3. Managers with higher abilities often have higher positions within a hierarchy (receiving greater reward for their work).

Hierarchies of Knowledge

This line of research proceeds from the assumption that the major task of any manager is to solve the problems arising during operation of an organization. The managers solve the problems owing to their knowledge and experience. The *specialization approach* (a specific worker concentrates only on definite classes of problems) is well-known to be efficient. At the same time, specialization leads to the problems of *coordination*, that is, choosing a worker being able to solve a specific problem. It turns out that an organization in the form of hierarchy represents rather efficient way of such coordination. The main problem of constructing a certain organization is to reach a trade-off between the efficiency of knowledge utilization and the costs of coordination.

Within the model [53] proposed by L. Garicano, successful implementation of a technological process (in addition to common production factors, e.g., materials, equipment, and investments) requires workers' knowledge as their ability to solve the problems.

Generally speaking, an arbitrary organization may bear no resemblance to a hierarchy if it does not rely on subordination and division of labor (e.g., production/management). Nevertheless, Garicano demonstrates that in an optimal hierarchy the workers are specialized either in production activity or in "problem solving." Only a single class is engaged in production (the corresponding employees are naturally said to be workers, while the other ones are called managers). Similar hierarchies of managers solving the problems were also studied by A. Beggs [13].

General conclusions of the considered line of research are as follows:

1. A primary task of managers consists in solving the problems that arise during operation of an organization;
2. A hierarchical control structure is an efficient way to organize the problem-solving process;
3. Management activity requires special skills, and benefits of dividing the employees into managers and workers depends on possible gains provided by specialization;
4. Lower levels of a hierarchy serve to exempt upper-level managers from "daily chores" and allow them to focus on solving more complicated strategic problems.

Multilevel Data Processing Hierarchies

An essential constraint of the previous models is the idea of hierarchy as a sequence of subordinate levels. An alternative approach to modeling of organizational hierarchies (eliminating this shortcoming) is developed in the works by R. Radner, T. van Zandt, and others (see [136] and the survey in [149]). The approach is based on an analogy between operation of organizational hierarchies and of computer networks. It claims that a primary function of managers in organizations lies in *data processing*.

It is assumed that the major characteristics of data processing are the total time required to evaluate a function and the number of processors. Both incur costs and should be improved. Consider a given number n of the elements being summed up and a fixed number P of processors. The problem is to build an *efficient* computer network (an organizational hierarchy) to minimize the total computation time. (See Figure 6.11.)

Meanwhile, in organizations data is commonly processed in the *systolic mode*, when separate data *cohorts* are supplied periodically. T. van Zandt has shown that an efficient hierarchy for the systolic mode consists of a set of optimal trees for processing a single cohort and a cohort routing mechanism balancing the workload of these trees.

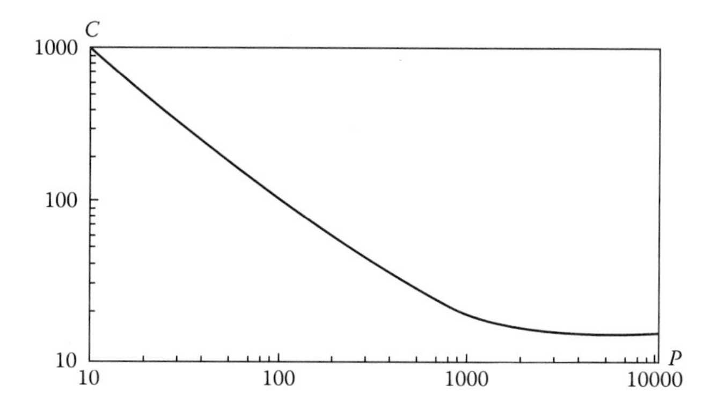

Figure 6.11 A relationship between the computational time and the number of processors for n = 10,000 (in logarithmic scale).

P. Bolton and M. Dewatripont [18] extended the approach by Radner and van Zandt. In their model, an organization operates in continuous time, and a new data cohort is available at any instant. The problem is summing up all data elements and accumulating them at top managers' "hands" with a specific frequency.

Let us outline general conclusions for this line of research:

1. If the main function of managers is data processing, the control system has a hierarchical structure to avoid duplication of managerial work.
2. Managers' headcount and the number of hierarchy levels grow as the organization gets expanded, and so does the response time of an organization to external shocks.
3. An efficient work allocation among managers implies leveling their information workload.
4. There is a tendency to separate hierarchical levels (interaction among managers at different levels is limited).

Hierarchies and the Theory of Teams

The basic model of the theory of teams (developed by J. Marshak and R. Radner [98]) represents an intermediate between classical optimization problem and game-theoretic models of conflict. A *team* of agents is considered maximizing a common efficiency criterion by choosing a vector of individual actions under incomplete information about the state of an environment. The problem is to coordinate the agents' decisions, that is, to suggest rational rules of choosing the actions under their personal beliefs.

J. Cremer [46] considered a firm composed of n production units (*factories*) manufacturing several types of products. Note the output of some factories can be the input for other factories; that is, technology of the firm requires *transfers* of products among the factories. In the elementary case (the demand for the products is known along with the cost functions of the factories), coordination aims to define the vectors of production and transfers for every factory; this is done for satisfying the demand with minimal manufacturing costs.

However, costless a posteriori transfer adjustment could be impossible for the whole firm. Consequently, one should partition the firm into smaller *blocks* consisting of one or several factories. Provided a fixed partitioning of the set of factories into blocks, the coordination problem is efficiently solved within each block, since factory cost functions and demands are supposed to be known within a block. The volumes of transfers among the blocks should be fixed in advance (prior to realization of random factors). Therefore, the problem of optimal organization is reduced to choosing the best admissible* partition of the factories into blocks.

The key idea of the model is the following. In the first place, the blocks should include the factories having strongest technological relations, since transfers between them are subject to the largest variations (under changes of random factors). The model is limited by the fact that it considers only organizations with a single intermediate control level (the one of

---

* Admissible partitions are limited by the maximum number of factories entering a block.

blocks). Introducing new intermediate levels in Cremer's model seems pointless.

In addition to the limited number of hierarchical levels, Cremer's model possesses another essential disadvantage. It implies that growing sizes of blocks improve the efficiency of the whole organizational structure. However, the time costs and the financial costs necessary to acquire true information on the factory cost functions in a large block are not considered. Yet J. Geanakoplos and P. Milgrom [55] suggested an advanced model to account for these costs in an explicit form. Thus, the span of control becomes an internal parameter of the model.

We formulate general conclusions for this line of research as follows:

1. A hierarchical control structure is a certain compromise between complicated coordination of numerous blocks and the costs to maintain intermediate control levels.
2. A complicated nature of coordination could be caused by incomplete information regarding relevant parameters of control blocks.
3. Larger blocks are made first from the factories having the strongest interrelations.

Decision-Making Hierarchies

First and foremost, control in an organization means making various decisions. Thus, the opinion that *decision making* represents the primary function of managers is very common. The quality of decisions finally determines the efficiency of the whole organization and organizational structure as well.

The model suggested by R. Sah and J. Stiglitz [139] states that an organization is engaged in analyzing the flow of *projects*. With a certain probability, each project could be either "good" (bringing profits) or "bad" (causing losses). The task of any organization is to choose good projects and implement them. The projects are assessed by managers. A separate manager can make mistakes by accepting bad projects and rejecting good ones. The following is proposed to improve the efficiency of such choice. The decision regarding implementation of a particular project should be made on the basis of the collective opinion of several managers. For this, different organizational structures can be formed such as a *committee*, a *hierarchy*, or a *polyarchy*.

These organizational forms could be combined to construct a more complicated organizational structure. For instance, one may consider a hierarchy of committees, a hierarchy of polyarchies, or a committee of hierarchies [140].

An alternative approach to modeling decision-making hierarchies is developed by O. Hart and J. Moore [70]. In their model a hierarchy of m identical managers represents a certain superstructure over a fixed set of *assets*. An asset is any object that could be used (separately and with other assets) to gain profits to an organization. Thus, any combination of assets is a potential project to be implemented by an organization.

Every manager can control a nonempty subset of assets $A_i \subseteq N$. With the probability $p(A_i)$ a manager may have an *idea* of how to use this subset. Being implemented, the idea yields the profits $v(A_i) > 0$. The role of a hierarchy lies in that the ideas of upper-level managers are implemented first. If they have no idea, the decision is made by their immediate subordinates. If the latter have nothing to suggest, the right is passed to their subordinates, and so on.

A hierarchy is chosen to maximize the expected profits, and the major problem consists in that the hierarchy must be constructed ex ante (i.e., before the managers actually have ideas). The general theoretical results derived by Hart and Moore enable substantial narrowing of the set of hierarchies being "suspiciously optimal." For instance, the researchers show the higher is the manager's level in an optimal hierarchy, the smaller is the probability that he or she has an idea. The authors provide a complete solution to the problem of optimal hierarchy in the case of two identical assets with an arbitrary number of managers.

Hierarchies and Contract Theory

Contract theory is a field of mathematical economics studying the *incentive* problems for economic agents in uncertain conditions (see Chapters 3–4 of this textbook).

D. Melumad et al. [102] used the classical *model of adverse selection* to investigate the influence of decentralized contracts on the efficiency of organizational operation. In particular, a system is considered consisting of two agents and a single principal, which yields the income from the agents' production output. The agent's production effort incurs costs that depend on individual characteristics of the agent, i.e., the agent's *type*. The latter represents private information being known only to the agent in question. Every agent is rewarded depending, first, on his or her effort level and, second, on his or her type reported (in general, agents manipulate and report untrue types).

Within the framework of the classical model of adverse selection, the principal directly interacts with both agents and offers rewards (*contracts*) to them; the contracts are based on the actions of the agents and the reported estimates of their types. The posed problem possesses a well-known solution (including the formula of average profits of the principal under optimal contracts).

This scheme of principal–agents interaction corresponds to a two-level hierarchy. Its efficiency must be compared with that of the decentralized scheme. (The principal concludes a contract only with agent 1 and delegates the right to conclude a subcontract with agent 2.) The surprising result is that under certain conditions decentralization does not bastardize the efficiency of the incentive mechanism. (Naturally, decentralization does not increase the efficiency.)

In their paper [102] E. Maskin, Y. Qian, and C. Xu use a rather simple model to compare the advantages of two popular organizational structures, notably, a *functional* or *unitary* model (a U-organization) and a *multidivisional* model (an M-organization). In a functional organization, units form large blocks according to similarity of their functions. In a divisional organization, units are combined in blocks according to the geographical location principle (self-sufficient divisions) or according to the product principle (independent product lines).

The models discussed in this paragraph compare a few simple hierarchies. The matter is that analyzing the incentive problems in uncertain conditions appears cumbersome and time-consuming.

Therefore, game-theoretic analysis of the problem of organizational hierarchy design implies the following:

1. The type of an organizational structure has an essential impact on the managers' preferences, as well as on their decisions.
2. An optimal structure essentially depends on the efficiency of a managers' incentive scheme.
3. The efficiency of an incentive scheme (in its turn) depends on the uncertainty level and on the possibilities to monitor the manager's work.
4. As a rule, contracts decentralization reduces the efficiency of an organization (though in some cases this can be avoided).

6.3 General Model of Management Hierarchy

Complicacy in organizational structure modelling attributes, first, to difficulties in formulating the corresponding mathematical model and, second, to difficulties in formal analysis of the model. From the above overview we see that all known mathematical models represent management activity in a simplified form. Provided the whole variety of management activity types, just one or two types are selected (e.g., coordinating subordinates, monitoring of their actions, problem solving, decision making, and data processing).

Challenges in formal analysis of the stated mathematical models are due to imperfect mathematical tools used to solve design problems for complex systems structure. The majority of these models allow for a unified mathematical statement in the form of a discrete optimization problem, yet different authors involve various (specific) mathematical approaches to analyze them.

The current section describes a unified framework for hierarchy optimization. (Almost all known models are reduced to it.) Also it proves a series of general conclusions about optimal organizational hierarchies.

Suppose that a finite set of workers N, a set of feasible hierarchies $\Omega \subseteq \Omega(N)$, and the *cost function* $C: \Omega \to [0; +\infty)$ are specified. The cost function maps a feasible hierarchy into a nonnegative real value. The optimal hierarchy problem is to find a feasible hierarchy having minimal costs:

$$H^* \in \operatorname*{Arg\,min}_{H \in \Omega} C(H).$$

Generally, the set of feasible hierarchies Ω can be a strict subset of the set of all hierarchies $\Omega(N)$ that control the set of workers N. In particular (and depending on interpretation of a specific problem), an optimal tree or an optimal r-hierarchy could be of interest.

Under a small number of workers, the problem in question is solved by the exhaustive search among all feasible hierarchies (evidently, in the general case this is the only method). However, as a rule there are a great many hierarchies; thus, defining the cost function by simply enumerating the values of all hierarchies that belong to the set Ω seems impossible. Consequently, the cost function is determined by an analytical expression or a certain algorithm depending on structural parameters of a hierarchy (e.g., the number of managers, the number of subordinates, and the tasks performed by the managers).

To design efficient methods of optimal hierarchy search, one should introduce assumptions concerning the cost function type (i.e., consider a specific class of functions). Such assumptions may base on an experimental research of the cost functions in real-world organizational hierarchies or on general economic considerations.

Sectional Cost Functions

Definition 6.9 [110]

Let the set of workers N be given. The manager's cost function is said to be *sectional* if it depends only on the groups of workers controlled by immediate subordinates of this manager.

Suppose that the manager m has r immediate subordinates $v_1, v_2, \dots, v_r$ within a hierarchy H. Then his or her costs could be written in the form

$$c(m, H) = c(s_H(v_1), \dots, s_H(v_r)).$$

The number of arguments in a sectional cost function equals the span of control (the number of immediate subordinates) of the manager, and the function is defined for all spans. The value of a sectional cost function remains the same for any permutation of its arguments (groups). Hence, a sectional cost function maps an arbitrary nonempty set of workers' groups into a number representing the costs of the manager that possesses immediate subordinates controlling these groups.

In the case of the sectional function, manager's costs are independent of how the work is performed inside the units headed by his or her immediate subordinates. Actually, the costs depend only on the groups of workers controlled by the immediate subordinates. For instance, the costs of the manager m (within the hierarchies shown in Figures 6.12a and

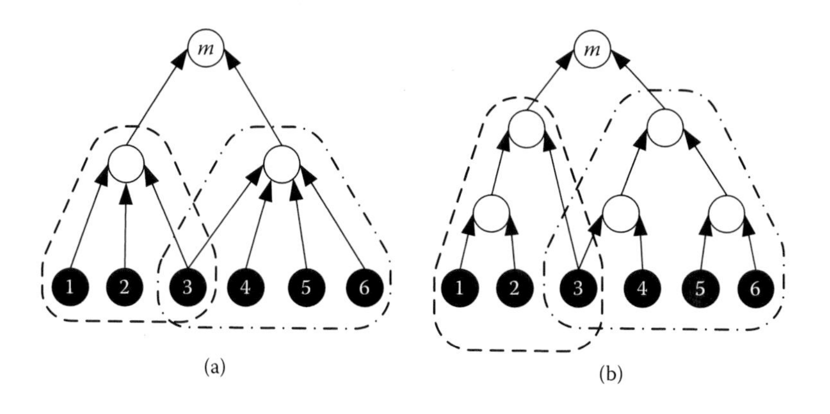

Figure 6.12 Illustrating the notion of sectional cost function.

6.12b) are identical. Indeed, in both hierarchies the manager m has two immediate subordinates controlling the groups {1, 2, 3} and {3, 4, 5, 6}. Note that the total costs of these hierarchies may differ.

Recall the model discussed in Section 6.1; manager's costs $c(\cdot)$ are defined by the technological flows and the function $\phi(\cdot)$. The internal and external flows of a manager depend only on the groups controlled by his or her immediate subordinates $v_1, \ldots, v_r$. The manager's cost function (6.4) depends just on the groups $s_H(v_1), \ldots, s_H(v_r)$ and, thus, is sectional. Therefore, *the model of management hierarchy built over a technological network considers a special case of a sectional cost function.*

Assertion 6.1 holds for the general model if a technical condition is imposed on a sectional function. The properties (i)–(iii) of optimal hierarchies (see Assertion 6.1) appreciably simplify evaluation of an optimal hierarchy; in addition, they imply the following. First, only finite sets of feasible hierarchies can be studied. Second, every manager would have (at least) two immediate subordinates. Even in this situation, the problem of finding an optimal hierarchy for an arbitrary sectional cost function appears rather intricate. Sometimes even such general formulation says a lot about the form of optimal hierarchy.

In the sequel, we consider a sufficient condition for optimality of a tree-shaped structure.

Definition 6.10 [151]

A sectional cost function of a manager is said to be *group-monotonic* if the costs of a manager do not decrease (1) by extending the groups controlled by immediate subordinates and (2) by adding new immediate subordinates to this manager. In other words, for any set of groups $s_1, \ldots, s_r$ the following inequalities take place:

$$c(s_1, s_2, \ldots, s_r) \le c(s, s_2, \ldots, s_r),$$

where the group s belongs to s_1 ($s_1 \subset s$);

$$c(s_1, s_2, \ldots, s_r) \le c(s_1, s_2, \ldots, s_r, s),$$

where s is an arbitrary group.

The group-monotonic property is explained in Figure 6.13. In particular, a section of a hierarchy subordinated to the

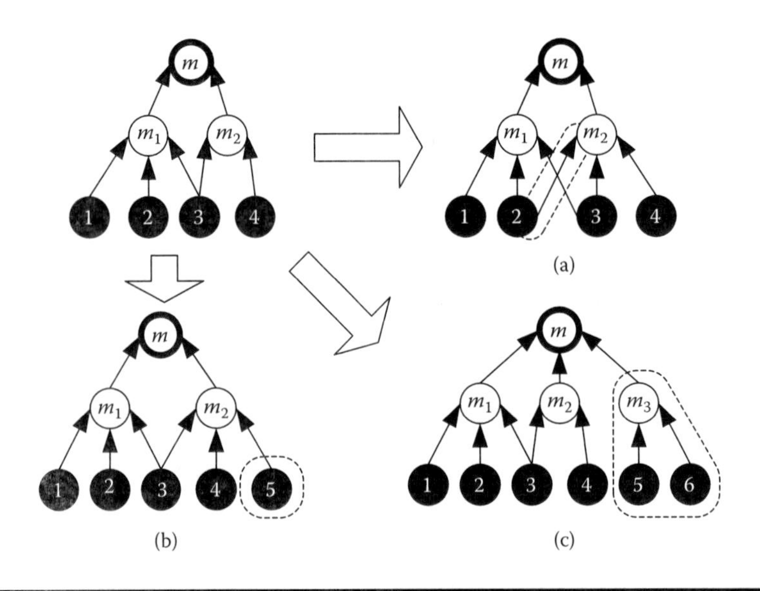

Figure 6.13 Illustrating the notion of group-monotonic cost functions.

manager m is demonstrated there; this manager has the immediate subordinates m_1, m_2.

The arrows correspond to possible ways of extending the groups controlled by immediate subordinates of the manager m (hierarchies 6.13a) and 6.13b) and to possible ways of adding new subordinate groups (hierarchy 6.13c). Hierarchy 6.13a is derived from the original one by extending the group subordinated to the manager m_2 (involving the subordinates of the manager m_1.) In hierarchy 6.13b the group subordinated to the manager m_2 is extended by adding new workers. Finally, in hierarchy 6.13c the manager m is given a new immediate subordinate (the manager m_3). Dashed lines indicate the added parts of the hierarchy. The manager's cost function is group-monotonic if for any similar transformations the costs of the manager m (see the heavy line in the figure) do not decrease.

Assertion 6.3 [111]

Assume that a cost function is group-monotonic. Then an optimal tree exists on the set of all hierarchies $\Omega(N)$ for a given set of workers N.

Notably, for a group-monotonic cost function an optimal hierarchy is tree-shaped.* Hence, numerical algorithms suggested in the literature [151] to find optimal trees† can be adopted for hierarchy search.

The group-monotonic property per se indicates nonoptimality of the *multiple subordination* of workers; if a cost function meets the property, then each worker (except the top manager) must have a single direct superior.

* The same result holds for hierarchy optimization on any feasible set Ω, which includes all trees, and on the set of r-hierarchies, which includes all r-trees.

† For an arbitrary sectional cost function, the exact algorithm to find an optimal tree possesses a high computational complexity. It provides a numerical solution being good-enough (in the sense of computational performance and accuracy) to the problem with at most 15–20 workers. Yet evaluation of an optimal nontree structure has an enormously higher complexity.

Optimality Conditions for Typical Hierarchies

We continue with consideration of conditions ensuring optimality of hierarchies with minimum and maximum possible spans of control.

Definition 6.11 [110]

A sectional cost function is called *narrowing* if for any manager m with the immediate subordinates $v_1, \ldots, v_r$ ($r \geq 3$) it is possible (1) to re-subordinate some of his or her workers to a new manager m_1 and (2) to make the manager m_1 directly subordinated to the manager m such that both operations do not increase the costs of the hierarchy. Accordingly, a sectional cost function is called *widening* if any resubordination does not decrease the costs of the hierarchy.

Figure 6.14 illustrates the definition introduced. In particular, the left-hand side (hierarchy (a)) represents a section of the manager m, composed of the manager and his or her immediate subordinates v_1, v_2, and v_3 (they could be either managers or workers). The right-hand side of the figure (hierarchy (b)) demonstrates the same section of the hierarchy after several immediate subordinates of the manager m (e.g., the workers v_1 and v_2) are resubordinated to the new manager m_1 (indicated by bold line). Such a reformation existing for any manager and not increasing the costs of the hierarchy, the cost function is

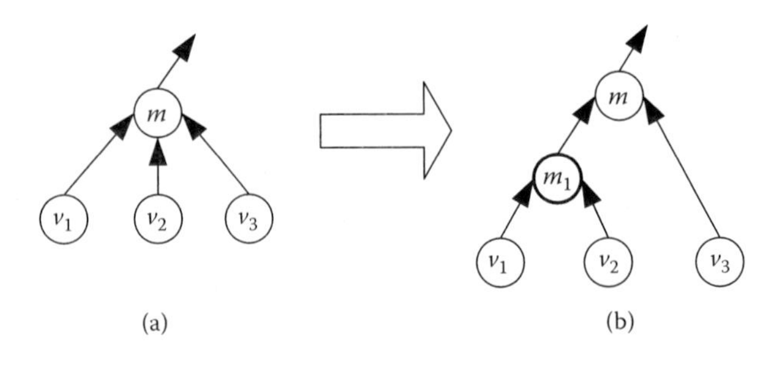

Figure 6.14 Illustrating the notions of the narrowing and the widening cost functions.

narrowing. When any reformation does not reduce the costs of the hierarchy, the cost function is widening.

We underline that Definition 6.11 requires nonincreasing or nondecreasing costs of the whole hierarchy. Variation of the total costs includes the costs of the manager m_1 being added and variation of the costs related to the manager m (after resubordination he or she has less immediate subordinates). Therefore, a narrowing cost function makes it necessary (at least) that the costs of the manager m do not grow as several of his or her immediate subordinates are substituted for the manager m_1.

In practice, the notion of narrowing function means the following. Consider a hierarchy where a certain manager possesses more than two immediate subordinates. It is always beneficial to hire an "assistant" for him or her (thus reducing the load on this manager). On the contrary, under a widening cost function, one always takes advantage of eliminating intermediate managers. These arguments are the key idea to prove the following result.

Assertion 6.4 [110, 111]

For a narrowing cost function over the set $\Omega(N)$, there exists an optimal 2-hierarchy. On the other hand, a fan structure is optimal for a widening cost function.

Consequently, in case of a narrowing cost function, one should look for an optimal hierarchy only among 2-hierarchies over the set $\Omega(N)$ (or, alternatively, over an arbitrary set Ω including all 2-hierarchies). At the same time, if the cost function is of the widening type and a fan hierarchy is feasible, exactly the latter appears optimal.

Suppose that a cost function is narrowing and group-monotonic. Using Assertions 6.3–6.4, one would easily show that an optimal hierarchy forms a 2-tree. Moreover, Definition 6.11 could be weakened for a cost function being group-monotonic. In this case, the conditions stated in the definition must be satisfied only when all workers $v_1, \ldots, v_r$ control nonintersecting groups of workers. If the weakened condition holds true, the corresponding cost function is said to be *narrowing over nonintersecting groups*. For a monotonous function, this property leads to the optimality of a fan hierarchy.

The result of Assertion 6.4 utilizes the monotonicity of the hierarchy costs with respect to sequential operations of resubordination. Each operation does not increase (does not decrease) the costs of the hierarchy for a narrowing (respectively, widening) function. And optimal hierarchies are exactly the ones that could not be transformed by any resubordination. Similarly, one may define other transformations for hierarchies and use the monotonicity of hierarchy costs.

For instance, assume it is necessary to find an optimal hierarchy over the set of feasible hierarchies $\Omega(N)$ (with a narrowing cost function). According to Assertion 6.4, an optimal hierarchy exists among 2-hierarchies. Now, imagine in a 2-hierarchy H a manager m possesses immediate subordinates m_1 and m_2. In addition, the former controls workers v and w, while the latter is a manager for workers v' and w' (see Figure 6.15a). The mentioned workers may have other superiors not shown in the figure. Denote by s_1 and s_2 the groups controlled by the managers m_1 and m_2, respectively.

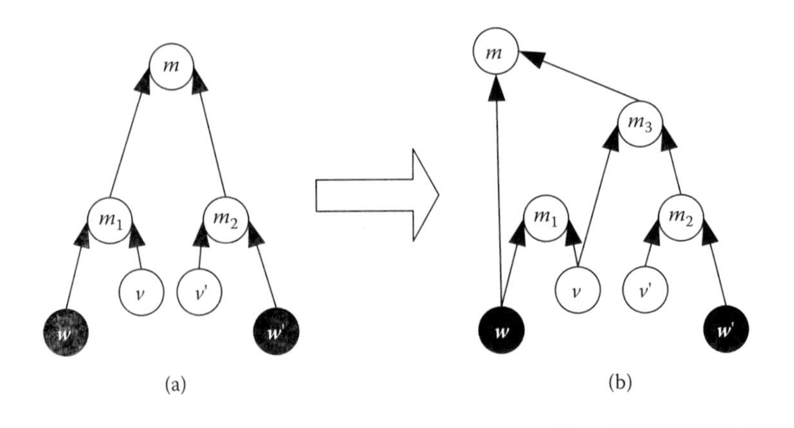

Figure 6.15 Illustrating the notion of strongly narrowing cost function.

Let us transform the illustrated section of the hierarchy as follows. Eliminate the links coming from the managers m_1 and m_2 to the manager m, add a new manager m_3 and make him subordinated to the manager m. Moreover, assign the worker v and the manager m_2 as immediate subordinates of the manager m_3; appoint the manager m a direct superior to the worker w (see Figure 6.15b). Evidently, such transformation does not modify the costs of the managers shown in the figure. At the same time, the costs of the whole hierarchy vary by the quantity

$$c(s_1 \setminus \{w\}, s_2) + c((s_1 \setminus \{w\}) \cup s_2, \{w\}) - c(s_1, s_2).$$

Exactly the same operation could be applied to manager m_2.

Definition 6.12 [110, 111]

A narrowing function is called *strongly narrowing* if for arbitrary groups s_1 and s_2 composed of two or more workers at least one of the following conditions is met:

1. For any $w \in s_1 : c(s_1, s_2) \geq c(s_1 \setminus \{w\}, s_2)$ $+ c((s_1 \setminus \{w\}) \cup s_2, \{w\})$.
2. For any $w \in s_2 : c(s_1, s_2) \geq c(s_1, s_2 \setminus \{w\})$ $+ c(s_1 \cup (s_2 \setminus \{w\}), \{w\})$.

Hence, for a strongly narrowing cost function the described transformation is always possible with no total costs increase in a hierarchy. Such transformation is impossible for a sequential hierarchy. Thus, the assertion follows.

Assertion 6.5 [110, 111]

For a strongly narrowing cost function there exists an optimal sequential hierarchy over the set $\Omega(N)$.

Therefore, provided a strongly narrowing cost function, an optimal hierarchy over $\Omega(N)$ is among sequential hierarchies; to find it, use analytical and numerical algorithms are developed.

It is easily shown* that the functions being group-monotonic (and the ones being not) may appear narrowing, widening, or even not to belong to these classes. Moreover, a function may be simultaneously narrowing and widening.

Generally, a sectional cost function $c(s_1, \ldots, s_r)$ of a manager is a function of sets, thus representing a rather complex mathematical object. In the general case, specifying a sectional function lies in exhaustive enumeration of the values for all feasible combinations of groups. As a rule, this seems impossible due to an overwhelming number of such combinations.

Elucidating the properties of sectional functions, let us rewrite the cost function of a manager in a compact form. For this, establish a relationship between each group (or a combination of groups) and a single (or several) numerical characteristics and assume that the cost function depends on these characteristics.

The simplest way is introducing a certain measure over the set of workers. Every worker $w \in N$ is assigned a positive real value $\mu(w)$ representing his or her *measure*. The measure $\mu(s)$ related to the group of workers $s \subseteq N$ is defined as the total measure of all workers belonging to the group, that is, $\mu(s) := \Sigma_{w \in s} \mu(w)$. Then we assume that the cost function of a manager may be expressed as the following function of $(r + 1)$ variables: $c(s_1, \ldots, s_r) = c(\mu_1, \ldots, \mu_r, \mu)$. Note that $\mu_1, \ldots, \mu_r$ here stand for the measures of the groups controlled by immediate subordinates of the manager, while μ is the measure of the group controlled by the manager himself. The described function is said to be *measure-dependent*.† In practice, the worker's measure can be interpreted, for instance, as the complexity level of the work performed. Consequently, the measure of the group corresponds to the total complexity or the total volume of work performed by the group (the cost of controlling the group depends on the complexity).

---

* The corresponding examples are given in [110].

† A manager's cost function is defined for any number of immediate subordinates; it is symmetric with respect to any permutation of the arguments $\mu_1, \ldots, \mu_r$.

Example 6.4

Suppose that all workers are identical and have unit measure. Then the measure of the group is equal to the number of workers in it. Accordingly, the cost function of a manager depends on the number of immediate subordinates and on the numbers of workers controlled by these immediate subordinates. •

Undoubtedly, specifying the measure of workers is not a unique (yet, the simplest) way to introduce numerical characteristics of the groups. In particular, we have earlier discussed "flow-type" manager's cost functions that depend on material flows, financial, information, and other flows among subordinate groups of workers. Let us give some examples of cost functions that depend on the measures.

Example 6.5

Imagine the manager's costs are proportional to the measure of the group controlled by him or her: $c(\mu_1, \ldots, \mu_r, \mu) = \mu$. In this case, the fan hierarchy is optimal among all feasible hierarchies. Indeed, according to the definition, any hierarchy includes a manager who controls the group N composed of all workers, and only in a fan hierarchy this manager is unique. However, optimal hierarchies are not as trivial in the class of r-hierarchies (where $r > 1$ is a constant). •

Example 6.6

Assume that the cost function of a manager depends on the number r of his or her immediate subordinates and on the measure μ of the group controlled by the manager. A special case is the *multiplicative* cost function $c(r, \mu) = \phi(r)\chi(\mu)$, with $\phi(\cdot)$ and $\chi(\cdot)$ designating nonnegative strictly increasing functions.

In a multiplicative function, the costs $\phi(r)$ related to working with immediate subordinates are multiplied by the "responsibility factor" $\chi(\mu)$, which depends on the measure of the group controlled by the manager. •

Example 6.7

Consider several more complicated measure-dependent manager's cost functions, first introduced in [110]:

(i) $c(\mu_1,\ldots,\mu_r,\mu) = [\mu_1^\alpha + \ldots + \mu_r^\alpha - \max(\mu_1^\alpha,\ldots,\mu_r^\alpha)]^\beta$,

(ii) $c(\mu_1,\ldots,\mu_r,\mu) = [\mu_1^\alpha + \ldots + \mu_r^\alpha]^\beta$,

(iii) $c(\mu_1,\ldots,\mu_r,\mu) = [\mu^\alpha / \max(\mu_1^\alpha,\ldots,\mu_r^\alpha) - 1]^\beta$,

(iv) $c(\mu_1,\ldots,\mu_r,\mu) = \left[\sum_{i=1}^{r} (\mu^\alpha - \mu_i^\alpha)\right]^\beta$.

Here α and β are certain nonnegative parameters used to adjust these functions subject to specific conditions. In the sequel, we will refer to these functions by their corresponding numbers (e.g., the function (i)).

The cost functions (i)–(iv) model the costs of a manager to implement different types of interaction (see the literature on management science) with his or her subordinates within a section.

First, suppose a *semileader* exists among immediate subordinates of a manager. This person performs all his or her duties and requires no additional attention from a direct superior. This type of interaction is modeled by the function (i); notably, the manager's costs are defined by complexity levels of the groups controlled by all immediate subordinates except that of the semileader. A subordinate controlling the largest group is supposed to be a semileader. When *no leader* presents among immediate subordinates, then the manager incurs the costs to control all immediate subordinates (as in the function (ii)).

Assume there is a leader among immediate subordinates of a manager within a section; he or she assists in solving interaction problems among other employees in a section (e.g., using his or her authority or experience). This leader reduces the costs of his or her boss. This type of interaction is modeled by the cost function (iii). The higher the complexity level for the group controlled by the leader, the higher his or her value in the view of the manager (the costs of the latter are better decreased).

The function (iv) may describe the costs *due to one-to-one communication with subordinates*. The costs depend on the differences between complexity level of the whole section and complexity levels of the groups controlled by section members.*

The following results are due to S. Mishin [110]. Evidently, the functions (i) and (ii) are group-monotonic (in contrast to the functions (iii) and (iv)). One could establish when these functions are narrowing or widening. The function (i) turns out widening under $\beta \leq 1$ and narrowing otherwise. Hence, $\beta \leq 1$ implies optimality of a fan hierarchy, while $\beta > 1$ makes optimal a certain 2-tree (see Figure 6.16a).

Similarly, the function (ii) is shown to be widening if $\beta \leq 1$; moreover, it appears widening over nonintersecting groups provided that $\beta > 1$ and $\alpha \geq 1$. In both cases, a fan structure is optimal (see Figure 6.16b). In the domain $\beta > 1$, $\alpha < 1$, the function (ii) is neither widening nor narrowing (even over nonintersecting combinations of groups). In other words, Assertion 6.4 no more assists in evaluating the optimal hierarchy in these cases. However, the function (ii) is group-monotonic, and the optimal hierarchy is provided by a tree (see Assertion 6.3).

It is shown that under $\beta \geq 1$ the functions (iii) and (iv) are narrowing; that is, the optimal hierarchy is provided by a 2-hierarchy with minimum costs (see Assertion 6.4). In the case $\beta < 1$, the tree ensuring the minimum costs could be obtained using specialized algorithms [151]. However, this tree could be a suboptimal hierarchy, since the functions (iii) and (iv) are not group-monotonic. •

Widening and narrowing cost functions lead to optimality of the extreme cases, viz. fan hierarchies and 2-hierarchies. As a rule, in real organizations an intermediate hierarchy takes

---

* For instance, consider a manager m with a subordinate group $s_H(m)$. Controlling an immediate subordinate m_1, the manager transmits to him or her information on a certain part of the group $s_H(m)$ not controlled by m_1. The volume of this information is determined by the difference between complexity levels $\mu(s_H(m))$ and $\mu(s_H(m_1))$. The total volume of information over all immediate subordinates gives the manager's costs (iv).

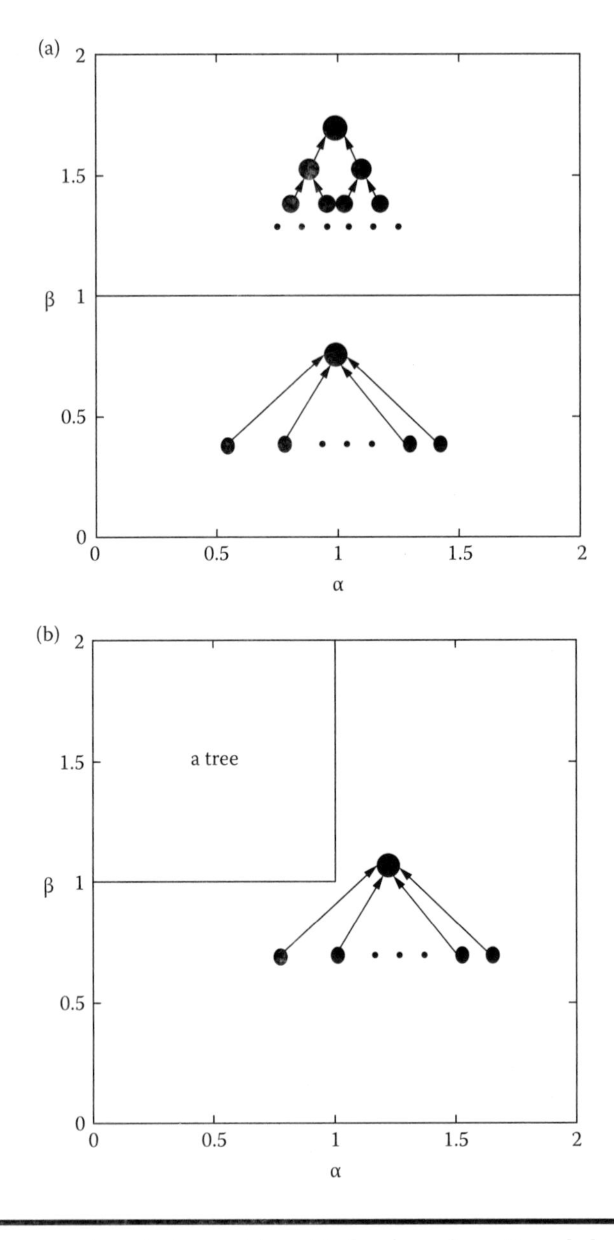

Figure 6.16 Optimal hierarchies: (a) the function (I) and (b) the function (II).

place with the span of control lying within $2 < r < +\infty$. Thus, a cost function describing such organization is neither widening nor narrowing. Hence, it seems important to design methods of solving the optimal hierarchy problem in this case. Till now, these methods have been proposed for the homogeneous cost functions.

6.4 Optimal Tree-Shaped Hierarchies

Homogeneous Cost Functions

Definition 6.13 [62, 151]

A measure-dependent cost function $c(\mu_1, \ldots, \mu_r, \mu)$ of a manager is called *homogeneous* if there exists a nonnegative number γ such that for any positive number A and any set of measures $\mu_1, \ldots, \mu_r, \mu$ the following identity holds:

$$c(A\mu_1, \ldots, A\mu_r, A\mu) = A^{\gamma} c(\mu_1, \ldots, \mu_r, \mu).$$

The number γ is known as the *degree of homogeneity* of the cost function.

Therefore, if the cost function is homogenous, its value grows by A^{γ} times when all its arguments are increased by A times.

Definition 6.14

An *r-dimensional simplex* D_r is a set composed of r-dimensional vectors $x = (x_1, \ldots, x_r)$ with nonnegative components such that $x_1 + \ldots + x_r = 1$. Elements of the simplex will be referred to as *r-proportions* or (simply) as *proportions*.

Obviously, for any manager possessing r immediate subordinates controlling the groups of measures $\mu_1, \ldots, \mu_r$, the vector $x := (\mu_1/\mu, \ldots, \mu_r/\mu)$ forms an r-proportion; here μ is the measure of the group controlled by the manager. We will say that the manager splits the subordinate group of workers among his or her immediate subordinates in the proportion x.

To find an optimal tree in the case of a measure-dependent cost function, one may use the existing numerical algorithms. Their analysis for different homogeneous cost functions allows for identifying a series of general properties of optimal trees. These properties are formalized in the following definition.

Definition 6.15 [62]

A tree is said to be (r, x)-*homogeneous* if each manager has exactly r immediate subordinates and splits among them the group of workers in the proportion $x = (x_1, \ldots, x_r)$. The number r is called a *span of control* of the uniform tree.

Example 6.8

Figure 6.17 shows three uniform trees; for each worker, the figure demonstrates the measure of the group controlled by the worker. Hierarchy (a) is a 3-tree with the proportion $x = (1/3, 1/3, 1/3)$. The tree appears symmetric (actually, trees are always symmetric if the workers possess identical measures). Hierarchy (b) makes a 2-tree with the proportion (1/2, 1/2), while hierarchy (c) represents a 2-tree with the proportion (1/3, 2/3). •

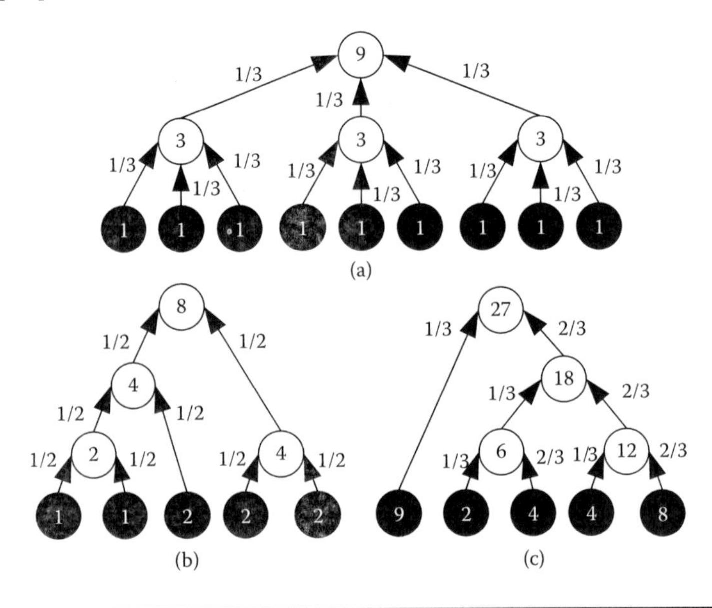

Figure 6.17 Examples of uniform trees.

As far as the problem considered is discrete, sometimes no uniform tree exists for the given set of workers (the only exception is a fan hierarchy being per se homogeneous). At the same time, if a uniform tree exists the corresponding costs are easily evaluated.

Assertion 6.6 [62]

Suppose that the set of workers $N = \{1, \ldots, n\}$ with the measures $\mu(1), \ldots, \mu(n)$ is given and the manager's cost function $c(\mu_1, \ldots, \mu_r)$ has the degree of homogeneity γ. If there exists a uniform tree H with the span of control r and the proportion $x = (x_1, \ldots, x_r)$, then the costs are defined by

$$C(H) = \begin{cases} \left|\mu^\gamma - \sum_{j=1}^{n} \mu(j)^\gamma\right| \dfrac{c(x_1, \ldots, x_r)}{\left|1 - \sum_{i=1}^{r} x_i^\gamma\right|}, & \text{for } \gamma \neq 1, \\ \left(\mu \ln \mu - \sum_{j=1}^{n} \mu(j) \ln \mu(j)\right) \dfrac{c(x_1, \ldots, x_r)}{-\sum_{i=1}^{r} x_i \ln x_i}, & \text{for } \gamma = 1. \end{cases} \tag{6.6}$$

Here $\mu := \mu(N) = \Sigma_{i=1}^{n} \mu(i)$ indicates the total measure of all workers.

A Lower-Bound Estimate of Optimal Tree Costs

One may use formula (6.6) to choose the *best uniform tree* from the set of all uniform trees. To find the attributes of such a tree, one should minimize (6.6) over all feasible spans of control r and proportions x. The pair (r, x), where the minimum is achieved, gives the parameters of the best uniform tree. Substitute them in formula (6.6) to obtain the corresponding costs.

Naturally, under a fixed set of workers $N = \{1, \ldots, n\}$ with the measures $\mu(1), \ldots, \mu(n)$, a top manager of any tree would have (at most) n immediate subordinates. Thus, to find the best uniform tree, it suffices to perform minimization over all r from 2 to n.

Moreover, each immediate subordinate of the top manager would control (at least) a single worker. Hence, the measure of the group controlled by him or her would be not smaller than the minimum measure among all workers. Consequently, every component x_i ($i = 1, \ldots, r$) of the proportion describing any uniform tree would be not smaller than

$$\varepsilon := \min_{i \in N} \mu(i) \Big/ \sum_{i \in N} \mu(i).$$

Consider an arbitrary nonnegative value ε; denote by $D_r(\varepsilon)$ a part of the simplex D_r such that each component is above or equal to ε.

Then for a fixed cost function the minimum costs of a uniform tree are determined by n and by the measures $\mu(1), \ldots, \mu(n)$ of the workers. These costs satisfy the following formula:

$$C_L(N) = \begin{cases} \left| \mu^{\gamma} - \sum_{j=1}^{n} \mu(j)^{\gamma} \right| \min\limits_{k=2\ldots n} \min\limits_{y \in D_k(\varepsilon)} \dfrac{c(y_1, \ldots, y_k)}{\left| 1 - \sum_{i=1}^{k} y_i^{\gamma} \right|}, \text{ when } \gamma \neq 1, \\ \left(\mu \ln \mu - \sum_{j=1}^{n} \mu(j) \ln \mu(j) \right) \min\limits_{k=2\ldots n} \min\limits_{y \in D_k(\varepsilon)} \dfrac{c(y_1, \ldots, y_k)}{-\sum_{i=1}^{k} y_i \ln y_i}, \\ \text{when } \gamma = 1, \end{cases} \tag{6.7}$$

where $\mu = \Sigma_{i \in N} \mu(i)$ and $\varepsilon = \min_{i \in N} \mu(i)/\mu$.*

It has been empirically confirmed that an optimal tree-shaped structure (over the set of all trees) "seeks" to be a

* Since the set $D_r(\varepsilon)$ is compact, the minima in (6.7) are attained under weak conditions imposed on the cost function. Just require it to be lower semicontinuous on the simplex. In the sequel, we assume that this condition holds.

uniform tree. Hence, it seems natural to assume that, if there exists the best optimal tree (with the span of control $r(n, \varepsilon)$ and the proportion $x(n, \varepsilon)$) for the given set of workers, this tree is optimal over the set of all trees. Actually, even a stronger result can be proved rigorously.

Assertion 6.7 [62]

Let the set of workers $N = \{1, \ldots, n\}$ with the measures $\mu(1), \ldots, \mu(n)$ and the homogeneous manager's cost function $c(\cdot)$ (with γ as the degree of homogeneity) be specified. Then the optimal tree costs are not smaller than $C_L(N)$. In other words, the function $C_L(N)$ provides a lower-bound estimate of optimal tree costs.

Imagine the problem of optimal r-tree is posed (find a certain tree such that every manager has at most r subordinates). Then a lower-bound estimate of its costs is equal to the costs of the *best uniform* r-tree, i.e., a uniform tree with the span of control not exceeding r.

Evidently, the costs of the best uniform r-tree are expressed by

$$C_L^r(N) = \begin{cases} \left|\mu^\gamma - \sum_{j=1}^n \mu(j)^\gamma\right| \min\limits_{k=2\ldots\min[n,r]} \min\limits_{y \in D_k(\varepsilon)} \dfrac{c(y_1, \ldots, y_k)}{\left|1 - \sum_{i=1}^k y_i^\gamma\right|}, \text{when } \gamma \neq 1, \\ \left(\mu \ln \mu - \sum_{j=1}^n \mu(j) \ln \mu(j)\right) \min\limits_{k=2\ldots\min[n,r]} \min\limits_{y \in D_k(\varepsilon)} \dfrac{c(y_1, \ldots, y_k)}{-\sum_{i=1}^k y_i \ln y_i}, \\ \text{when } \gamma = 1. \end{cases} \tag{6.8}$$

Therefore, the following statement is true.

Assertion 6.8 [62]

Within the conditions of Assertion 6.7, the costs of the optimal r-tree are not smaller than $C_L^r(N)$.

The derived lower-bound estimate of optimal tree costs has numerous applications. For instance, under a large number of workers the estimate serves as a good approximation of optimal tree costs (for the details see [64]).

The Model of a Problem-Solving Hierarchy

Consider a control system—a hierarchy (in the sense of Definition 6.1) over the set of workers $N = \{1, \ldots, n\}$.

Suppose that problem solving is the sole function of the manager. Then his or her workload is proportional to the number of decisions made. The decisions are to solve the problems being faced by the subordinates of the manager. Given a workload P (the number of decisions per unit time), the manager's costs are equal to P^β, where β is the elasticity coefficient. Marginal costs do not decrease with the workload (i.e., $\beta \geq 1$). The elasticity β reflects manager's competence. Compared with low-skilled managers (with greater value of β), highly skilled ones (with lower β) require smaller costs to solve the same number of problems. And vice versa, under the same costs, the latter solve more problems than the former.

A manager makes decisions based on the reports provided by his or her immediate subordinates. Assume that the volume of a report (prepared by a certain subordinate for his or her superior) makes μ^α. The parameter μ means the measure of the group controlled by the subordinate. Moreover, suppose that the number of decisions made by a superior is proportional to the total volume of reports received by him.

The parameter $\alpha \in [0,1]$ is interpreted as a rate of data compression in a report. This rate depends on how typical the problems faced by workers are. Notably, if many workers have similar problems, the volume of the corresponding report has a small dependence on the number of workers. Thus, α

is appreciably smaller than 1 ($\alpha = 1$ means every problem is unique and data compression is impossible).

Therefore, if k immediate subordinates of a manager control groups with the measures $\mu_1, \ldots, \mu_k$, then the total volume of their report is $\mu_1^\alpha + \ldots + \mu_k^\alpha$; the manager's costs are equal to (within a constant)

$$c(\mu_1, \ldots, \mu_k) = (\mu_1^\alpha + \ldots + \mu_k^\alpha)^\beta. \qquad (6.9)$$

Construction of an optimal organizational hierarchy is reduced to finding a hierarchy admitting the minimum total costs of the managers. In addition to optimal hierarchy evaluation, it appears interesting to analyze the relation between its basic characteristics (e.g., span of control and costs) and parameters of the model (the level α of problems' uniqueness and managers' competence β).

Such analysis enables, first, selecting efficient policy to reduce management expenses and, second, planning reorganization as a response to changes in external conditions.

For the current example, the expression (6.9) defining the manager's costs coincides with the formula of the cost function (ii) introduced in Example 6.7. This cost function is group-monotonic; hence, an optimal hierarchy forms a tree. Figure 6.16b makes it obvious that under $\alpha \geq 1$ or $\beta \leq 1$ a fan hierarchy appears optimal. Thus, it seems curious to find an optimal tree in the domain $\alpha < 1$, $\beta > 1$. Solving this problem requires evaluating the parameters of the best uniform tree (the span of control and the proportion).

The degree of homogeneity of the cost function is $\alpha\beta$. Suppose $\alpha\beta \neq 1$. Formula (6.7) implies that to find the best proportion (for a fixed span of control k) one has to obtain a proportion $(y_1, \ldots, y_k)$ minimizing the function

$$\frac{(y_1^\alpha + \ldots + y_k^\alpha)^\beta}{\left|1 - \sum_{i=1}^{k} y_i^{\alpha\beta}\right|}. \qquad (6.10)$$

This minimization has been performed numerically. For the most important domain ($\alpha \in [0, 1]$, $\beta \in [1, 6]$), it has been shown that the best uniform trees are symmetric. Note that, for $\beta > 6.7$, there exist certain domains of the parameters α and β, where asymmetric proportions appear optimal. Optimal proportion being known, formula (6.7) serves to evaluate the best span of control of the uniform tree. The corresponding results of numerical computation are illustrated in Figure 6.18. Obviously, 2-trees are optimal for parameter β being high enough. Here this domain is marked by "2," while the dashed line indicates the subdomain where an optimal 2-tree is asymmetric. As soon as β decreases and α tends to the unity, 3-trees, 4-trees,... become optimal in a consecutive order (see the numbers "3," "4,"... in the figure).

Figure 6.18 demonstrates that the span of control increases as the manager's competence goes down (β decreases). That is, better managers possess more immediate subordinates. This is explicable in practice; indeed, highly skilled managers

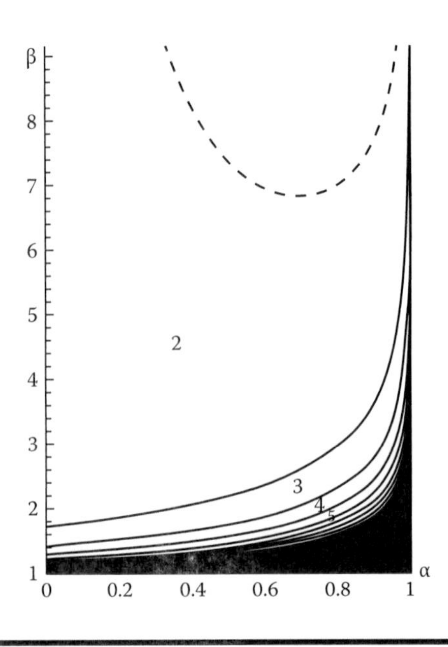

Figure 6.18 Spans of control in the best uniform tree for the cost function (II).

have greater workload against their low-skilled colleagues. An unexpected result consists in that the optimal span of control increases as the level of the problems uniqueness (the parameter α) goes up and data compression capabilities decrease.

Really, suppose that the measures of all workers exceed the unity. Apparently, growing α increases manager's workload (recall the expression $\mu^{\alpha} r^{1-\alpha}$). Hence, the same tendency is observed for the manager's costs. An increase in the span of control r enhances the volume of work performed by the manager in question.

The total number of managers within a uniform hierarchy makes $(n - 1)/(r - 1)$; that is, increasing the span of control reduces the number of managers. Such strategy is the "cheapest" way to deal with less typical problems. Otherwise, the hierarchy is getting more complicated, and solving numerous problems involves more managers (thus, increasing the total costs).

Let us estimate the impact exerted by α and β on the optimal hierarchy costs. The top manager's costs are defined by

$$\mu(N)^{\alpha\beta} c\left(\frac{1}{r(\alpha,\beta)}, \ldots, \frac{1}{r(\alpha,\beta)}\right) = \mu(N)^{\alpha\beta} r(\alpha,\beta)^{\beta(1-\alpha)}.$$

Naturally, as α (the level of problem uniqueness) increases, the costs of an optimal hierarchy and the top manager costs both grow. The optimal hierarchy costs strictly decrease as the managers' competence is improved (β goes down).

However, the relationship between the top manager costs and the parameter β is not so clear. Figure 6.19 demonstrates that growing ability (β goes down) gives a drop of the top manager costs (as his or her ability also grows). Then the costs start rising. The matter is that (according to the aforesaid) higher-skilled managers are endowed with more subordinates and the number of managers within the hierarchy reduces (thus, increasing the workload and even the cost of any manager).

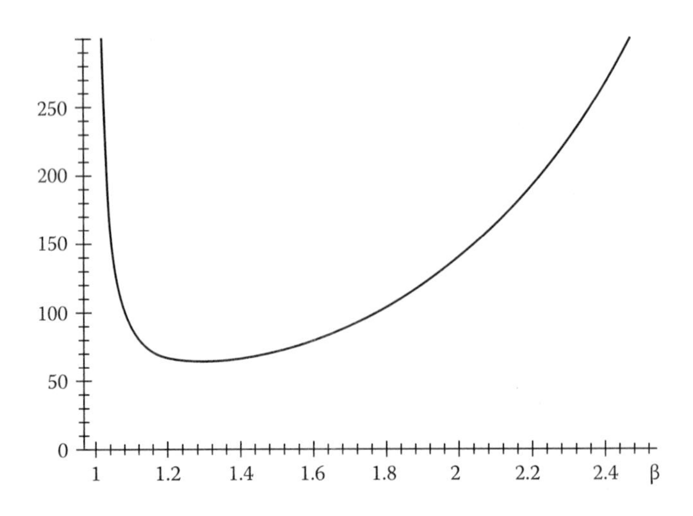

Figure 6.19 An example relationship between the top manager costs and managers' inverse ability β under $\alpha = 0.2$.

Therefore, if the top management of an organization invests in improving managers' ability (e.g., training), these actions lead to management expenses reduction. Nevertheless, the top management costs may even go up (of course, if the hierarchy is immediately adapted to new conditions).

Orders Execution and Plans Specification

In the previous model, information on the existing problems in a hierarchy moves bottom-up (from workers to a top manager). In addition to these ascending flows, descending information flows always present in an organization, coming downwards (from a top manager through his or her subordinates and to production workers). For instance, such information flows occur in the process of organization and operational planning or formulation and execution of orders. Consider a statement of the model in terms of order execution (planning processes are described by analogy).

Let n workers at the lower level be engaged in a business process of an organization. The tasks assigned to workers may require different efforts to control them. Hence, for each worker $w \in N = \{1, \ldots, n\}$, define a value $\mu(w)$ (a measure) describing the complexity of control of this worker. Then the volume of the most detailed instruction regulating the task for a group of workers $s \subseteq N$ (the volume of the document seems a reasonable metrics) would be proportional to the total measure $\mu(s)$ of workers entering the group. In other words, it would be proportional to the number of workers within the group.

At the same time, the manager controlling the group of the measure μ receives detailed instructions for the subordinate group, the volume of instructions constitutes μ^{α}, where $\alpha \in [0, 1]$ is the rate of data compression.

The manager's task consists, first, of analyzing every item of the order and, second, selecting immediate subordinates (there are k of them) that bear a relation to this piece of instruction. In fact, the manager should solve the problem of classification.

In the general case, the volume of work is defined by a function $\rho(k)$. The total manager workload P is proportional to $\mu^{\alpha}\rho(k)$.*

The manager's costs may have a nonlinear dependence on the workload P. Assume it represents a power function P^{β}. If the manager's work is just analyzing items of the order, his or her costs are given by $\mu^{\alpha\beta}\rho(k)^{\beta}$ (the multiplicative cost function in Example 6.6).

However, within the model considered, the manager should supplement and specify the obtained instructions, transforming them into k sets of instructions for his or her immediate subordinates. Suppose the required efforts are proportional to the extra volume $\mu_1^{\alpha} + \ldots + \mu_k^{\alpha} - \mu^{\alpha}$ of the detailed orders compared to the initial order.

* Such a relationship among the stated parameters is common not only for classification problems. For instance, work of a manager may include making his or her immediate subordinates familiar with items of an order. Imagine the manager invites all subordinates to a joint meeting; then his or her effort is proportional to the volume μ^{α} of the order. If each of k subordinates meets the manager privately, the manager's effort is proportional to $k\mu^{\alpha}$.

Hence, if k immediate subordinates of the manager control the groups of workers with the measures $\mu_1, \ldots, \mu_k$ (and the manager controls the group of the measure $\mu = \mu_1 + \ldots + \mu_k$), then his or her costs are defined by

$$(A\, \mu^\alpha \rho(k) + \mu_1^\alpha + \ldots + \mu_k^\alpha - \mu^\alpha)^\beta. \qquad (6.11)$$

In (6.11), A is a relative weight of a unitary analysis work against a unitary order specification work.

Similarly to the previous model, solve the problem of evaluating a hierarchy with the minimum managers' maintenance costs. The major interest lies in relationships among the span of control in an optimal hierarchy, its costs and the parameters of the model.

Note that reducing β corresponds to improving the total skills of the managers (as administrative workers) and to increasing their information processing capabilities. The rise of the parameter α could be interpreted as growing level, first, of the managers' specialization and, second, of their awareness regarding technological features of organization operation. Thus, they can prepare detailed orders for the subordinates.

Consider an arbitrary manager controlling the group of workers with the measure μ; suppose that r immediate subordinates of the manager control the groups of workers of the measures $\mu_1, \ldots, \mu_r$. Evidently, the volume $\mu_1^\alpha + \ldots + \mu_k^\alpha - \mu^\alpha$ of manager's order specification work possesses a non-monotonous dependence on α. In particular, it goes up until a specific value of α, and vanishes as α tends to the unity.*

The discussed nonmonotonicity follows from two tendencies. The first one is that the total volume of work for all managers within a hierarchy decreases as their level of specialization α increases (managers with higher specialization

* This cost function has the degree of homogeneity $\alpha\beta$. According to the accumulated statistics, the degree of homogeneity for the manager's costs in firms does not exceed 0.4.

make correct decisions faster). The second tendency lies in that the volume of work (related to specification of orders) increases accordingly; that is, managers with higher specialization provide better specification to the orders.

In practice, it is often difficult to recruit a sufficient number of workers who are experts in technology and management issues. Thus, one should try to reach a reasonable compromise. Study the impact of α and β on the optimal hierarchy costs to make a proper choice between technology experts and professional executives.

The cost function (6.11) is group-monotonic; hence, an optimal hierarchy could be found in the class of trees. The best uniform tree is definitely symmetric if we consider the most interesting range of parameters. Thus, let us confine ourselves to optimal symmetric trees in this case.

Formula (6.7) implies that (under fixed parameters α, β, and A) finding the span of control of the best uniform tree requires evaluating an integer number $k > 1$ minimizing the function

$$\frac{(A\,\rho(k) + k^{1-\alpha} - 1)^{\beta}}{|1 - k^{1-\alpha\beta}|}.$$

This is a standard minimization problem being easily solved numerically. For instance, Figure 6.20 illustrates optimal spans of control in the case $A = 0.5$ (recall this parameter describes work content to analyze the order against its specification). The figure shows that optimal span of control increases under reducing the level of the manager's specialization (α goes down) or under improving their skills (β goes down).

At the same time, this rule could be violated if plan specification requires more time than the analysis does (i.e., if A is decreased).

For instance, Figure 6.21 demonstrates optimal spans of control in the case $A = 0.05$. Evidently, under small values of β (higher managerial skills) the same relationship between span

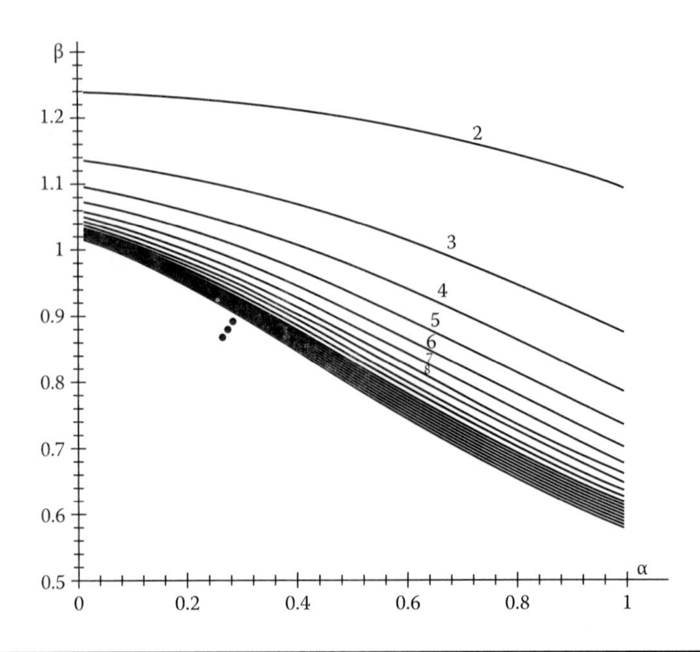

Figure 6.20 Optimal spans of control for the cost function (6.11) under $A = 0.5$.

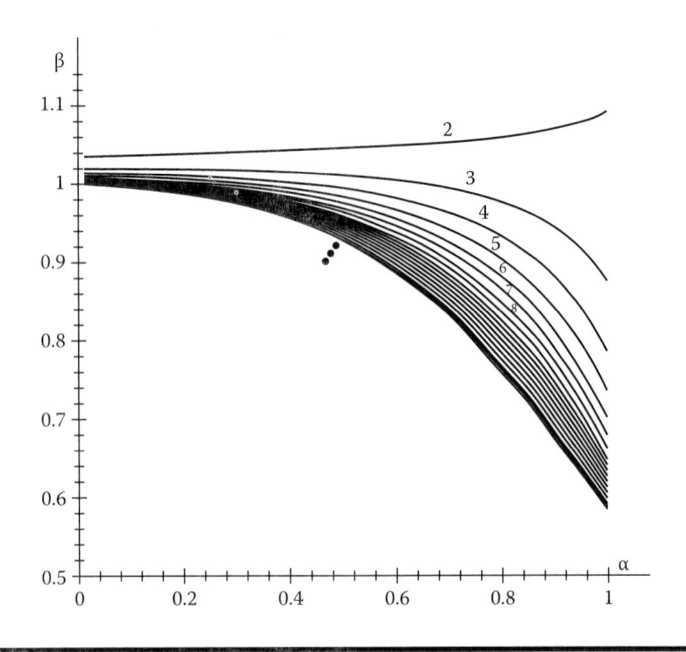

Figure 6.21 Optimal spans of control for the cost function (6.11) under $A = 0.05$.

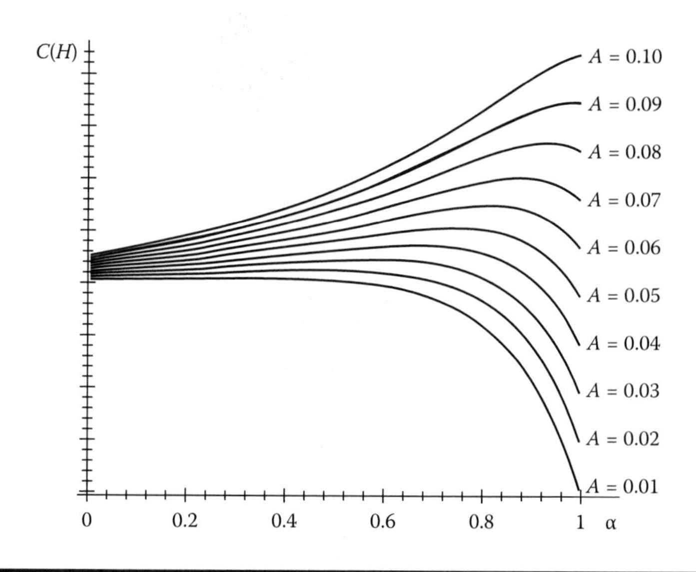

Figure 6.22 An example of the relationship between the optimal hierarchy costs and the level of managers' competence α under n = 1,000.

of control and the specialization level α takes place. In contrast, for $\beta > 1.03$ the optimal span of control grows as α goes up.

The family of curves in Figure 6.22 corresponds to the relationships between the hierarchy maintenance costs and the parameter α under different values of A. In fact, the figure clarifies the following. If order specification prevails in the managers' work (i.e., the parameter A is small), then the hierarchical costs are minimum under a large level of the managers' specialization (maximum value of α). Hence, in this case, an organization benefits from managers being experts in technology. However, a growing role of work related to classifying the items of the order (A goes up) results in increasing costs of specialized experts; for $A > 0.05$, the hierarchical costs are minimized by the minimum value of α. This means it is beneficial to form an organizational hierarchy from "versatile" managers.

Still, concise conclusions regarding advantages of certain management actions (for modifying an organizational structure) require a detailed analysis of a concrete situation in an organization.

Control Costs and the Size of a Firm

In the view of mathematical economics, it is really important to know how the costs of a hierarchical control system depend on the size of a firm. This relationship being available, one would answer the following questions of principle (remember a series of models surveyed in the previous section): Can a hierarchically controlled firm grow infinitely? Do boundaries of a firm exist, being unprofitable to exceed? If they do, further growth of the economics seems possible only by interaction of independent economic agents (at a competitive market).

The problem could be illustrated using the following elementary model. Let a function $V(n)$ describe the revenue of a firm depending on the number of workers n being engaged in production process. A reasonable assumption lies in that this function is non-decreasing with respect to n. For simplicity, suppose that the revenue $V(n)$ obeys constant returns to scale and is written as $p\ n$.

Imagine the employees' wages constitute the only sort of expenses of a firm. If all workers have an identical wage σ, the wage fund makes $\sigma\ n$.

However, we know that production workers per se are not enough for a normal operation of a firm. Management efforts are also an essential input. So, a hierarchy of managers that require additional costs is needed. Under a given number of workers n, there exists an optimal hierarchy of managers with the minimum possible costs $C(n)$.

Then the profit of the organization (the revenue minus the costs) is defined by $(p - \sigma)\ n - C(n)$; this formula implies the following. If the costs $C(n)$ of an optimal hierarchy grow linearly for large n (the corresponding rate of growth is less than $p - \sigma$), then the profits increase with respect to n. In other words, an infinite growth of the firm provides definite benefits. Now, assume that under large n the costs of a hierarchy grow superlinearly. In this case, there is an optimal number of workers n^* such that exceeding it reduces the profit of the firm. Provided a convex function $C(n)$, the optimal number in question is evaluated by the condition $C'(n^*) = p - \sigma$. This means further growth of the firm seems unprofitable.

There is exactly a linear dependence between the cost of the management hierarchy and the size of the firm that initiated a long-lasting dispute in the economic literature.

Using homogeneous cost functions (depending on the parameters of the model), one may describe both types of the above relationships between the optimal costs of a hierarchy and the size of an organization (see Figure 6.24).

Suppose the degree of homogeneity γ (related to the cost function of a manager within an organizational hierarchy) is not equal to the unity. Then the optimal hierarchy costs grow proportionally to $|n - n^\gamma|$. In other words, with $\gamma < 1$ the hierarchy costs grow proportionally to n for large n, and such a firm allows for an infinite growth. On the contrary, under $\gamma \geq 1$ the costs of a management hierarchy grow superlinearly (are proportional to n^γ). Consequently, there exists a limit size being unprofitable to exceed. (See Figure 6.23.)

To settle the matter regarding possibility of infinite growth of a firm, we should know whether the degree of homogeneity of the manager's cost function* exceeds unity or not. For a specific organization, the degree of homogeneity of the cost function could be roughly estimated according to the following procedure. Assume that the manager's cost function within a hierarchy is homogeneous, and the existing organizational structure is optimal. In this case, if the costs to keep a manager in the hierarchy are greater than the total costs to keep all his or her immediate subordinates, then the degree of

---

* In addition to wages, the manager's costs may include overheads required to organize his or her work (e.g., office rents and machinery). The wages of secretaries and assistants could be also included.

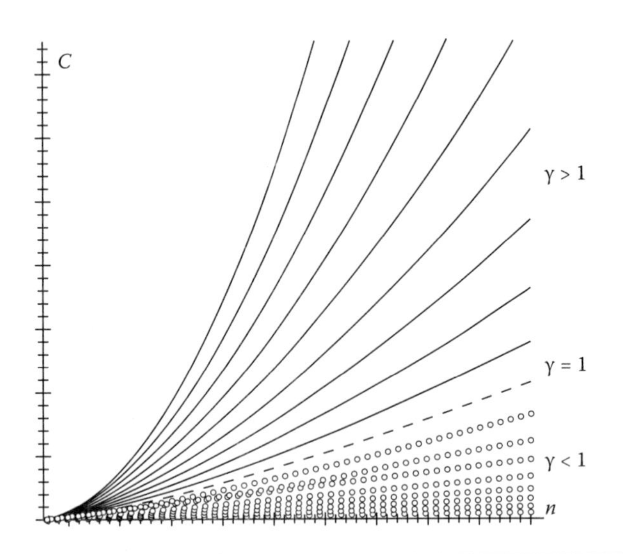

Figure 6.23 An example of the relationship between the optimal hierarchy costs *C* and the size of organization *n* (for different degrees of homogeneity).

homogeneity of the cost function exceeds the unity. Otherwise, the degree of homogeneity appears less than the unity.

Therefore, roughly speaking, if maintaining a manager within an organization is cheaper than maintaining his or her immediate subordinates taken together, then the firm can grow infinitely. If not, the firm is limited to a certain size before it becomes unprofitable.

TASKS AND EXERCISES

6.1*. Give the definitions and illustrative examples for the following terms:
Organizational design
Worker
Flow function
Flow intensity
Technological network
Production line
(Direct) superior

(Direct) subordinate
Group
Subordinate group of workers
Tree
Span of control
Hierarchy
  Sequential hierarchy
  Fan-type hierarchy
  r-hierarchy
  Optimal hierarchy
Flow
  External flow
  Internal flow
  Manager flow
Cost function
  Sectional cost function
  Group-monotonic cost function
  Narrowing cost function
  Measure-dependent cost function
  Homogeneous cost function

6.2. What graphs illustrated in Figure 6.24 are hierarchies? Managers and workers are marked with white and black circles, respectively.

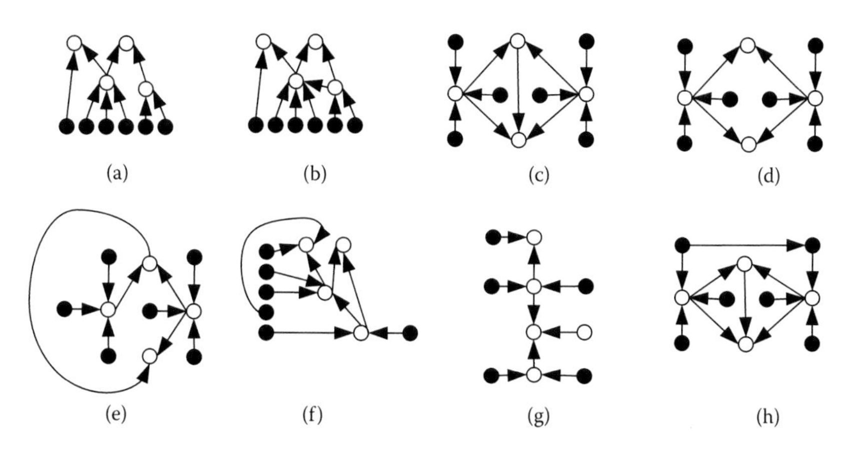

Figure 6.24 Examples of graphs for organizational structures.

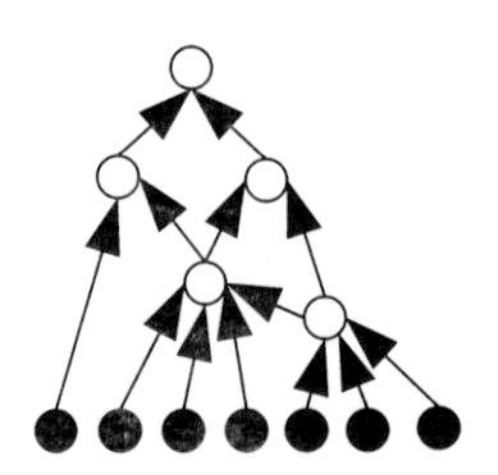

Figure 6.25 An example of an organizational hierarchy.

6.3. What properties of hierarchies (see Assertion 6.1) are violated by the graphs in Figure 6.24 representing nonhierarchical structures?

6.4. Using Assertion 6.1, explain why the hierarchy shown in Figure 6.25 should not be considered as a candidate for an optimal hierarchy. Find a hierarchy possessing smaller or the same costs.

6.5. Check whether the hierarchy demonstrated by Figure 6.25 represents a 3-hierarchy.

6.6. Prove Lemma 6.1 using the notion of subordinate group of workers and the acyclic property of a hierarchy (see [110] for help).

6.7*. Prove Lemma 6.2.

6.8. Consider the model of control hierarchy built over a technological graph. Prove that the managers together control all technological flows, that is, for any hierarchy H:

$$\sum_{m \in M} F_H^{int}(m) = \sum_{w, w' \in N} f(w, w').$$

6.9. Consider a technological network with single-component flows (see Figure 6.26). Using the results of Section 6.1, find an optimal hierarchy and the corresponding costs if the manager's cost function is $\phi(x) = 10 + x + (x + 10)^{0.5}$.

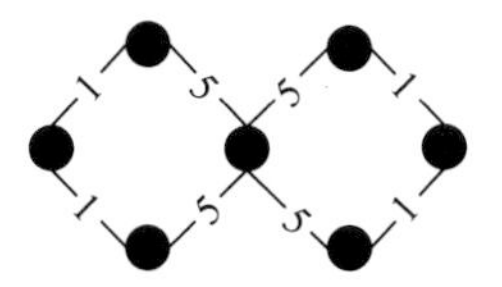

Figure 6.26 An example of a flow network.

6.10. Demonstrate that the multiplicative cost function $c(r, \mu) = \phi(r)\chi(\mu)$ introduced in Example 6.6 is group-monotonic if the functions $\phi(r)$ and $\chi(\mu)$ are nondecreasing.

6.11. Show that for any nonnegative numbers α and β the cost function (ii) introduced in Example 6.7 is group-monotonic (see [110] for assistance).

6.12*. Show that for any nonnegative numbers α and β the cost function (i) introduced in Example 6.7 is group-monotonic (see [110] for assistance).

6.13*. Give an example illustrating that the group-monotonic property is violated for the cost function (III) introduced in Example 6.7 (see [110] for assistance).

6.14*. Give an example illustrating that the group-monotonic property is violated for the cost function (IV) introduced in Example 6.7 (see [110] for assistance).

6.15. Demonstrate that the cost function (i) introduced in Example 6.7 is narrowing (see Definition 6.11) for $\beta > 1$ (see [110] for assistance).*

6.16. Demonstrate that the cost function (iv) introduced in Example 6.7 is narrowing (see Definition 6.11) for $\beta > 1$ (see [110] for assistance).

6.17*. Demonstrate that the cost function (i) introduced in Example 6.7 is strongly narrowing (see Definition 6.12) for $\alpha\,\beta > 1$, $\beta > 1$ (see [110] for assistance).

---

* Use the following form of the Minkowski inequality. For arbitrary nonnegative numbers $x_1, \ldots, x_k$ and any number $\gamma \geq 1$: $(x_1 + \ldots + x_k)^\gamma \geq x_1^\gamma + \ldots + x_k^\gamma$.

6.18. Assume that all workers possess unit measures. Derive a formula for optimal costs of a hierarchy under the cost function (i) where $\alpha\,\beta > 1$, $\beta > 1$ (use the results of Task 6.17 and Assertion 6.5, and see [110] for further assistance).

6.19*. Demonstrate that the cost function (ii) introduced in Example 6.7 is widening (see Definition 6.11) for $\beta \le 1$ (see [110] for assistance).

6.20*. Demonstrate that the cost function (ii) introduced in Example 6.7 is widening over nonintersecting groups (see the corresponding definition after Assertion 6.4) for $\alpha \ge 1$, $\beta > 1$ (see [110] for assistance).

6.21*. Demonstrate that for the homogeneous multiplicative cost function $c(r, \mu) = \mu^{\gamma}\phi(r)$ the best uniform tree is symmetric; that is, the minimum with respect to y in formula (6.7) is achieved at $y = (1/k, \ldots, 1/k)$ (see [62] for assistance).

6.22. Find the span of control of the best symmetric uniform tree for the following cost functions of a manager (see more details in [62]):
(a) $c(r, \mu) = \mu^{0.5} r$,
(b) $c(r, \mu) = r^2$,
(c) $c(r, \mu) = \mu^{0.5} r^{0.5}$,
(d) $c(r, \mu) = \mu\, r$.

Conclusion

In this textbook we have considered in brief the methods and models of control in organizational systems. The more detailed exposition can be found in [127]. Nowadays the results of control theory and mechanism design are widely applied to solve management problems in various areas [19, 124] in corporate management, in project and multi-project management, in ergatic systems (the ones combining a human, technology, and nature), and even in technical systems (being the main tool of control in multiagent systems).

One of the main challenges is the problem of control mechanisms combination and integration. Possible approaches and some applications are discussed in [19].

This book is just an introductory course to the world of control theory and mechanism design in organizations, an approach based on the rich theory and having very diverse applications. In different chapters of this book we employed a number of theoretical concepts and constructions, which were explained very concisely (e.g., utility function, strategic behavior, planning, project scheduling, auctions, and labor contracts), although each of them is a theme in the extensive literature. The next section contains a brief list of the topics and key words that may serve as a clue to the detailed discussion on the subjects related to the general line of the present book.

Topics for Self-Study

1. Systems and models
2. Operations research and control in organizations
3. Problem of identification in models of organizations
4. Modeling in micro- and macroeconomics
5. Simulation and business games
6. Integrated assessment in complex systems
7. Expert assessment in decision making
8. Multiple-criteria decision making
9. Reflexion in decision making
10. Organizational mechanisms in project management
11. Utility theory
12. Theory of choice
13. Preference relations
14. Subjectivity in decision making
15. Non-cooperative games
16. Cooperative games
17. Repeated games
18. Hierarchical games
19. Reflexive games
20. Bounded rationality
21. Fuzzy sets in models of organization
22. Models of collective behavior
23. Models of coordination and decentralization
24. Basic incentive schemes
25. Staff control in organizations

26. Structure control in organizations (multilevel organizations)
27. Institutional control in organizations
28. Informational control in organizations
29. Control in dynamic organizations
30. Control in multiagent organizations
31. Control in organizations with distributed control (common agency models)
32. Control in organizations with incomplete information
33. Control in organizations with joint activity constraints
34. Control in organizations with coalitional interactions among participants
35. Control models for organizations with unfair behavior of participants
36. Financing mechanisms
37. Models and methods of intrafirm control
38. Stable solutions of control problems in organizational systems
39. Models and methods of education management
40. Organizations with information reporting (planning mechanisms)
41. Incentive problems and labor supply models
42. Organizational mechanisms of project management
43. Optimization of organizational structure
44. Incentive schemes in multilevel hierarchies
45. Interconnection of compensation policy and the firm structure
46. Basic organizational structures
47. Multilevel organizational hierarchies
48. Teams and organizations
49. Dynamics of organizational structures
50. Network and hierarchical organizational structures
51. Hierarchical games and organizational structures

Literature

[1] Ackoff, R. *Classic Writings on Management*, 2nd ed. New York: Wiley, 1999. 356 pp.

[2] Aizerman, M., Aleskerov, F. *Theory of Choice*. Amsterdam: Elsevier, 1995. 324 pp.

[3] Armstrong, M. *Reward Management*. London: Cogan Page, 2000. 804 pp.

[4] Arrow, K. *Social Choice and Individual Values*. Chicago: University of Chicago Press, 1951. 204 pp.

[5] Ashby, W. *An Introduction to Cybernetics*. London: Chapman and Hall, 1956. 295 pp.

[6] Aumann, R. "Agreeing to Disagree." *The Annals of Statistics*. 1976. Vol. 4. No. 6. pp. 1236–1239.

[7] Aumann, R., Heifetz, A. "Incomplete Information." In R. Aumann and S. Hart, eds., *Handbook of Game Theory with Economic Applications*, Vol. III. Elsevier, 2002. pp. 1665–1686.

[8] Avdeev, V., Burkov, V., Enaleev, A. "Multichannel Active Systems." *Automation and Remote Control*. 1990. Vol. 51. No. 11. pp. 1547–1554.

[9] Azariadis, C. "Implicit Contracts and Underemployment Equilibria." *Journal of Political Economy*. 1975. No. 6. pp. 1183–1202.

[10] Baily, M. "Wages and Employment under Uncertain Demand." *Review of Economic Studies*. 1974. Vol. 41. No. 125. pp. 37–50.

[11] Baron, R., Greenberg, J. *Behavior in Organizations*. 9th ed. Upper Saddle River, NJ: Pearson Education Inc., 2008. 775 pp.

[12] Beckmann, M. "Some Aspects of Returns to Scale in Business Administration." *The Quarterly Journal of Economics*. 1960. Vol. 74. No. 3. pp. 464–471.

[13] Beggs, A. "Queues and Hierarchies." *The Review of Economic Studies*. 2001. Vol. 68. No. 2. pp. 297–322.
[14] Berge, C. *The Theory of Graphs and its Applications*. New York: John Wiley & Sons, Inc., 1962. 247 pp.
[15] Bernheim, D., Whinston, M. "Common Agency." *Econometrica*. 1986. Vol. 54. No. 4. pp. 923–942.
[16] Bertalanffy, L. *General System Theory: Foundations, Development, Applications*. New York: George Braziller, 1968. 296 pp.
[17] Bolton, P., Dewatripont, M. *Contract Theory*. Cambridge, MA: MIT Press, 2005. 740 pp.
[18] Bolton, P., Dewatripont, M. "The Firm as a Communication Network." *The Quarterly Journal of Economics*. 1994. Vol. 109. No. 4. pp. 809–839.
[19] Burkov, V., Goubko, M., Kondrat'ev, V., Korgin, N., Novikov, D. *Mechanism Design and Management: Mathematical Methods for Smart Organizations (for managers, academics and students)*. New York: Nova Publishers, 2013. 204 pp.
[20] Burkov, V. *Foundations of Mathematical Theory of Active Systems*. Moscow: Nauka, 1977. 255 pp. (in Russian)
[21] Burkov, V., Danev B., Enaleev A., Naneva, T. "Competition Mechanisms for Allocation of Scarce Resources." *Automation and Remote Control*. 1988. Vol. 49. No. 11. pp. 1505–1513.
[22] Burkov, V., Enaleev, A. "Stimulation and Decision-making in the Active Systems Theory: Review of Problems and New Results." *Mathematical Social Sciences*. 1994. Vol. 27. pp. 271–291.
[23] Burkov, V., Enaleev, A. "Optimality of the Principle of Fair Play Management. Necessary and Sufficient Conditions for Reliability of Information in Active Systems." *Automation and Remote Control*. 1985. Vol. 46. No. 3. pp. 341–348.
[24] Burkov, V., Enaleev, A., Kondrat'ev, V. "Two-Level Active systems. IV. The Cost of Decentralization of Operating Mechanisms." *Automation and Remote Control*. 1980. Vol. 41. No. 2. pp. 829–835.
[25] Burkov, V., Enaleev, A., Novikov, D. "Operation Mechanisms of Social Economic Systems with Information Generalization." *Automation and Remote Control*. 1996. Vol. 57. No. 3. pp. 305–321.

[26] Burkov, V., Enaleev, A., Novikov, D. "Stimulation Mechanisms in Probability Models of Socioeconomic Systems." *Automation and Remote Control.* 1993. Vol. 54. No. 11. pp. 1575–1598.

[27] Burkov, V., Guevski, V., Naneva, T., Opoitsev, V. "Allocation of Water Resources." *Automation and Remote Control.* 1980. Vol. 41. No. 1. pp. 64–71.

[28] Burkov, V., Iskakov, M., Korgin, N. "Application of Generalized Median Voter Schemes to Designing Strategy-proof Mechanisms of Multicriteria Active Expertise." *Automation and Remote Control.* 2010. Vol. 71. No. 8. pp. 1681–1694.

[29] Burkov, V., Khohlov, V. "Cost-Efficient Taxation Mechanisms." *Automation and Remote Control.* 1992. Vol. 53. No. 5. pp. 703–709.

[30] Burkov, V., Kondrat'ev, V. "Two-Level Active Systems. I. Basic Concepts and Definitions." *Automation and Remote Control.* 1977. Vol. 38. No. 6. pp. 827–833.

[31] Burkov, V., Kondrat'ev, V. "Two-Level Active Systems. II. Analysis and Synthesis of Operating Mechanisms." *Automation and Remote Control.* 1977. Vol. 38. No. 7. pp. 997–1003.

[32] Burkov, V., Kondrat'ev, V. "Two-Level Active Systems. III. Equilibria in Above-Board Control Laws." *Automation and Remote Control.* 1977. Vol. 38. No. 9. pp. 1339–1347.

[33] Burkov, V., Kuznetsov, N., Novikov, D. "Control Mechanisms in Network Structures." *Automation and Remote Control.* 2002. Vol. 63. No. 12. pp. 1947–1965.

[34] Burkov, V., Lerner, A. "Fairplay in Control of Active Systems." In H. W. Kuhn and G. P. Szegö, eds., *Differential Games and Related Topics.* Amsterdam: North-Holland Publishing Company, 1971. pp. 164–168.

[35] Burkov, V., Novikov, D., Petrakov, S. "Mechanism Design in Economies with Private Goods: Truthtelling and Feasible Message Sets." *Systems Science.* 1999. Vol. 25. No. 1. pp. 71–78.

[36] Gubanov, D., Korgin, N., Novikov, D., Raikov, A. *E-Expertise: Modern Collective Intelligence.* Studies in Computational Intelligence, Vol. 558. Berlin: Springer-Verlag, 2014. 112 pp.

[37] Calvo, G. A., Wellisz, S. "Supervision, Loss of Control and the Optimal Size of the Firm." *The Journal of Political Economy.* 1978. Vol. 86. No. 5. pp. 943–952.

[38] Camerer, C. *Behavioral Game Theory: Experiments in Strategic Interactions*. Princeton: Princeton University Press, 2003. 544 pp.

[39] Chkhartishvili, A. "Bayes–Nash Equilibrium: Infinite-Depth Point Information Structures." *Automation and Remote Control*. 2003. Vol. 64. No. 12. pp. 1922–1927.

[40] Chkhartishvili, A., Novikov, D. "Information Equilibrium: Punctual Structures of Information Distribution." *Automation and Remote Control*. 2003. Vol. 64. No. 10. pp. 1609–1619.

[41] Chkhartishvili, A., Novikov, D. "Models of Reflexive Decision-Making." *Systems Science*. 2004. Vol. 30. No. 2. pp. 45–59.

[42] Chkhartishvili, A., Novikov, D. "Stability of Information Equilibrium in Reflexive Games." *Automation and Remote Control*. 2005. Vol. 66. No. 3. pp. 441–448.

[43] Cialdini, R. *Influence: Science and Practice*, 4th ed. Boston: Allyn & Bacon, 2000. 262 pp.

[44] Clarke, E. "Multipart Pricing of Public Goods." *Public Choice*. 1971. Vol. 11. No. 1. pp. 17–33.

[45] Coase, R. "The Nature of the Firm." *Economica*, New Series. 1937. Vol. 4. No. 16. pp. 386–405.

[46] Cremer, J. "A Partial Theory of the Optimal Organization of a Bureaucracy." *The Bell Journal of Economics*. 1980. Vol. 11. No. 2. pp. 683–693.

[47] Danilov, V., Sotskov, A. *Social Choice Mechanisms*. Berlin: Springer-Verlag, 2002. 191 pp.

[48] Dasgupta, P., Hammond, P., Maskin, E. "The Implementation of Social Choice Rules: Some General Results on Incentive Compatibility." *Review of Economic Studies*. 1979. Vol. 46. No. 2. pp. 185–216.

[49] D'Aspermont, C., Gerard-Varet, L. "Incentives and Incomplete Information." *Journal of Public Economics*. 1979. Vol. 11. No. 1. pp. 25–45.

[50] Drucker, P. *The Effective Executive: The Definitive Guide to Getting the Right Things Done*. New York: Collins Business, 2006. 208 pp.

[51] Fishburn, P. *Utility Theory for Decision Making*. New York: Wiley, 1970. 234 pp.

[52] Fudenberg, D., Tirole, J. *Game Theory*. Cambridge, MA: MIT Press, 1995. 579 pp.

[53] Garicano, L. "Hierarchies and Organization of Knowledge in Production." *The Journal of Political Economy*. 2000. Vol. 108. No. 5. pp. 874–904.
[54] Geanakoplos, J. "Common Knowledge." In R. Aumann and S. Hart, eds., *Handbook of Game Theory with Economic Applications*, Vol. 2. Amsterdam: Elseiver, 1994. pp. 1438–1496.
[55] Geanakoplos, J., Milgrom, P. "A Theory of Hierarchies Based on Limited Managerial Attention." *The Journal of Japanese and International Economies*. 1991. Vol. 5. No. 3. pp. 205–225.
[56] Germeier, Yu. *Non-antagonistic Games*. Dordrecht: D. Reidel Pub. Co., 1986. 331 pp.
[57] Germeier, Yu., Ereshko, F. "Auxiliary payments in games with a fixed sequence of moves." *USSR Computational Mathematics and Mathematical Physics*. 1974. Vol. 14. No. 6. pp. 1437–1450.
[58] Germeier, Yu. "Two Person Games with Fixed Sequence of Moves." *Soviet Mathematics*. 1971. Vol. 198. No. 5. pp. 1001–1004.
[59] Gibbard, A. "Manipulation of Voting Schemes: a General Result." *Econometrica*. 1978. Vol. 46. No. 3. pp. 595–616.
[60] Gordon, D. "A Neo-Classical Theory of Keynesian Unemployment." *Economic Inquiry*. 1974. No. 12. pp. 431–459.
[61] Goubko, M., Karavaev, A. "Coordination of Interests in the Matrix Control Structures." *Automation and Remote Control*. 2001. Vol. 62. No. 10. pp. 1658–1672.
[62] Goubko, M. "The Search for Optimal Organizational Hierarchies with Homogeneous Manager Cost Functions." *Automation and Remote Control*. 2008. Vol. 69. No. 1. pp. 89–104.
[63] Goubko, M. "Mathematical Models of Formation of Rational Organizational Hierarchies." *Automation and Remote Control*. 2008. Vol. 69. No. 9. pp. 1552–1575.
[64] Goubko, M. "Algorithms to Construct Suboptimal Organization Hierarchies." *Automation and Remote Control*. 2009. Vol. 70. No. 1. pp. 147–162.
[65] Grossman, S., Hart, O. "An Analysis of the Principal-Agent Problem." *Econometrica*. 1983. Vol. 51. No. 1. pp. 7–45.
[66] Groves, T. "Incentives in Teams." *Econometrica*. 1973. Vol. 41. No. 4. pp. 617–631.

[67] Harary, F. *Graph Theory*. Boston: Addison-Wesley Publishing Company, 1969. 214 pp.

[68] Harsanyi, J. "Games with Incomplete Information Played by 'Bayesian' Players." *Management Science*. Part I: 1967. Vol. 14. No. 3. pp. 159–182. Part II: 1968. Vol. 14. No. 5. pp. 320–334. Part III: 1968. Vol. 14. No. 7. pp. 486–502.

[69] Hart, O. "Optimal Labor Contracts under Asymmetric Information: an Introduction." *Review of Economic Studies*. 1983. Vol. 50. No. 1. pp. 3–35.

[70] Hart, O., Moore, J. "On the Design of Hierarchies: Coordination vs. Specialization." *The Journal of Political Economy*. 2005. Vol. 113. pp. 675–702.

[71] Hart, O., Holmstrom, B. "Theory of contracts." In T. F. Bewley, ed., *Advances in Economic Theory: Fifth World Congress*. Cambridge, UK: Cambridge University Press, 1987. pp. 71–155.

[72] Hillier, F., Lieberman, G. *Introduction to Operations Research*, 8th ed. Boston: McGraw-Hill, 2005. 1061 pp.

[73] Herzberg, F., Mausner, B. *The Motivation to Work*. New Brunswick: Transaction Publishers, 1993. 180 pp.

[74] Howard, N. "General Metagames: an Extension of the Metagame Concept." In A. Rapoport, ed., *Game Theory as a Theory of Conflict Resolution*. Dordrecht: Reidel, 1974. pp. 258–280.

[75] Howard N. "Theory of Meta-Games." *General Systems*. 1966. No. 11. pp. 187–200.

[76] Hurwicz, L. "Optimality and Informational Efficiency in Resource Allocation Processes." In K. Arrow, S. Karlin, and P. Suppes, eds., *Mathematical Methods in the Social Sciences*. Stanford, CA: Stanford University Press, 1960. pp. 27–46.

[77] Hurwicz, L. "On Informationally Decentralized Systems." In C. B. McGuire, R. Radner, and K. J. Arrow, eds., *Decision and Organization*. Amsterdam: North-Holland Press, 1972. pp. 297–336.

[78] Itoh, H. "Incentives to Help in Multi-agent Situations." *Econometrica*. 1991. Vol. 59. No. 3. pp. 611–636.

[79] Kahneman, D., Sclovic, P., Tversky, A. *Judgment under Uncertainty: Heuristics and Biases*. Cambridge, UK: Cambridge University Press, 1982. 544 pp.

[80] Kaufman, A. *Introduction to Fuzzy Arithmetic*. New York: Van Nostrand Reinhold Company, 1991. 384 pp.

[81] Keren, M., Levhari, D. "The Internal Organization of the Firm and the Shape of Average Costs." *The Bell Journal of Economics*. 1983. Vol. 14. No. 2. pp. 474–486.
[82] Kononenko, A. "On Multi-step Conflicts with Information Exchange." *USSR Computational Mathematics and Mathematical Physics*. 1977. Vol. 17. No. 4. pp. 104–113.
[83] Korgin, N. "Incentive Problems and Exchange Schemes." *Automation and Remote Control*. 2001. Vol. 62. No. 10. pp. 1673–1679.
[84] Kozielecki, J. *Psychological Decision Theory*. London: Springer, 1982. 424 pp.
[85] Krishna, V. *Auction Theory*, 2nd ed. New York: Academic Press, 2009. 336 pp.
[86] Kukushkin, N. "The Role of the Mutual Information of the Sides in Games of Two Players with Non-antagonistic Interests." *USSR Computational Mathematics and Mathematical Physics*. 1972. Vol. 12. No. 4. pp. 231–239.
[87] Laffont, G., Martimort, D. *The Theory of Incentives: The Principal-Agent Model*. Princeton: Princeton University Press, 2001. 421 pp.
[88] Lefebvre, V. *Algebra of Conscience*, 2nd ed. Berlin: Springer, 2010. 372 pp.
[89] Lefebvre, V. "Basic Ideas of the Reflexive Games Logic." In *Proc. Problems of Systems and Structures Researches*. Moscow: USSR Academy of Science, 1965. (in Russian)
[90] Lefebvre, V. *Lectures on the Reflexive Games Theory*. New York: Leaf & Oaks Publishers, 2010. 220 pp.
[91] Lefebvre, V. *The Structure of Awareness: Toward a Symbolic Language of Human Reflexion*. New York: Sage Publications, 1977. 199 pp.
[92] Leontjev, A. *Activity, Consciousness and Personality*. Englewood Cliffs, NJ: Prentice-Hall, 1978. 192 pp.
[93] Lewis, D. *Convention: A Philosophical Study*. Cambridge, MA: Harvard University Press, 1969. 232 pp.
[94] Lock, D. *The Essentials of Project Management*. New York: John Wiley & Sons, 2007. 218 pp.
[95] Ma, C. "Unique Implementation of Incentive Contracts with Many Agents." *Review of Economic Studies*. 1988. Vol. 55. No. 184. pp. 555–572.

[96] Macho-Stadler, I., Perez-Castrillo, J. D. "Centralized and Decentralized Contracts in a Moral Hazard Environment." *The Journal of Industrial Economics*. 1998. Vol. 46. No. 4. pp. 489–510.

[97] Mansour, Y. *Computational Game Theory*. Tel Aviv: Tel Aviv University, 2003. 150 pp.

[98] Marchak, J., Radner, R. *Economic Theory of Teams*. New Haven: Yale University Press, 1976. 345 pp.

[99] Mas-Collel, A., Whinston, M. D., Green, J. R. *Microeconomic Theory*. New York: Oxford University Press, 1995. 981 pp.

[100] Maskin, E. "Nash Equilibrium and Welfare Optimality." *The Review of Economic Studies*. 1977 (published 1999). Vol. 66. No. 1. pp. 23–38.

[101] Maskin, E., Qian, Y., Xu, C. "Incentives, Information and Organizational Form." *The Review of Economic Studies*. 2000. Vol. 67. No. 2. pp. 359–378.

[102] Melumad, D., Mookherjee, D., Reichelstein, S. "Hierarchical Decentralization of Incentive Contracts." *The RAND Journal of Economics*. 1995. Vol. 26. No. 4. pp. 654–672.

[103] Menard, C. *Institutions, Contracts and Organizations: Perspectives from New Institutional Economics*. Northampton: Edward Elgar Pub, 2000. 458 pp.

[104] Mertens, J., Zamir, S. "Formulation of Bayesian Analysis for Games with Incomplete Information." *International Journal of Game Theory*. 1985. No. 14. pp. 1–29.

[105] Mesarović, M., Mako, D., Takahara, Y. *Theory of Hierarchical Multilevel Systems*. New York: Academic, 1970. 294 pp.

[106] Meskon, M., Albert, M., Khedouri, F. *Management*, 3rd ed. New York: HarperCollins College Publishing, 1998. 777 pp.

[107] Milgrom, P., Roberts, J. *The Economics, Organization and Management*. Englewood Cliffs, NJ: Prentice Hall, 1992. 621 pp.

[108] Milgrom, P. *Putting Auction Theory to Work*. Cambridge, UK: Cambridge University Press, 2004. 368 pp.

[109] Mintzberg, H. *Structure in Fives: Designing Effective Organizations*. Englewood Cliffs, NJ: Prentice-Hall, 1983. 312 pp.

[110] Mishin, S. *Optimal Hierarchies in Firms*. Moscow: Institute of Control Sciences, 2005. 164 pp.

[111] Mishin, S. "Optimal Organizational Hierarchies in Firms." *Journal of Business Economics and Management*. 2007. No. 2. pp. 79–99.

[112] Mookherjee, D. "Optimal Incentive Schemes with Many Agents." *Review of Economic Studies*. 1984. Vol. 51. No. 2. pp. 433–446.

[113] Moulin, H. *Cooperative Microeconomics: A Game-Theoretic Introduction*. Princeton: Princeton University Press, 1995. 440 pp.

[114] Moulin, H. "Dominance Solvable Voting Schemes." *Econometrica*. 1979. Vol. 47. No. 6. pp. 1337–1351.

[115] Moulin, H. *Game Theory for Social Sciences*. New York: New York Press, 1986. 228 pp.

[116] Myerson, R. *Game Theory: Analysis of Conflict*. London: Harvard University Press, 1991. 568 pp.

[117] Myerson, R. "Incentive Compatibility and the Bargaining Problem." *Econometrica*. 1979. Vol. 47. No. 1. pp. 61–74.

[118] Myerson, R. "Optimal Auction Design." *Mathematics of Operations Research*. 1981. Vol. 6. No. 1. pp. 58–73.

[119] Myerson, R. "Optimal Coordination Mechanisms in Generalized Principal-Agent Problems." *Journal of Mathematical Economy*. 1982. Vol. 10. No. 1. pp. 67–81.

[120] Nash, J. "Equilibrium Points in N-person Games." *Proceedings of National Academy of Science*. 1950. Vol. 36. pp. 48–49.

[121] Neumann, J., Morgenstern, O. *Theory of Games and Economic Behavior*. Princeton: Princeton University Press, 1944. 776 pp.

[122] Nitzan, S. *Collective Preference and Choice*. Cambridge, UK: Cambridge University Press, 2010. 274 pp.

[123] Novikov, A., Novikov, D. *Research Methodology: from Philosophy of Science to Research Design*. Amsterdam: CRC Press, 2013. 200 pp.

[124] Novikov, D., Chkhartishvili, A. *Reflexion and Control: Mathematical Models*. Amsterdam: CRC Press, 2014. 298 pp.

[125] Novikov, D. *Control Methodology*. New York: Nova Science Publishing, 2013. 115 pp.

[126] Novikov, D. "Optimality of Coordinated Incentive Mechanisms." *Automation and Remote Control*. 1997. Vol. 58. No. 3. pp. 459–464.

[127] Novikov, D.A. *Theory of Control in Organizations*. New York: Nova Science Publishing, 2013. 358 pp.

[128] Novikov, D., Tsvetkov A. "Decomposition of the Game of Agents in Problems of Incentive." *Automation and Remote Control.* 2001. Vol. 62. No. 2. pp. 317–322.

[129] Novikov, D., Tsvetkov A. "Aggregation of Information in Incentive Models." *Automation and Remote Control.* 2001. Vol. 62. No. 4. pp. 617–623.

[130] Ore, O. *Theory of Graphs.* Providence: American Mathematical Society, 1972. 270 pp.

[131] Orlovsky, S. "Decision-making with a Fuzzy Preference Relation." *Fuzzy Sets and Systems.* 1978. Vol. 1. No. 3. pp. 155–167.

[132] Owen, G. *Game Theory.* Philadelphia: W.B. Saunders Company, 1969. 228 pp.

[133] Peregudov, F., Tarasenko, F. *Introduction to Systems Analysis.* Columbus, OH: Glencoe/McGraw-Hill, 1993. 324 pp.

[134] Perlman, R. *Labor Theory.* New York: Wiley, 1969. 237 pp.

[135] Qian, Y. "Incentives and Loss of Control in an Optimal Hierarchy." *The Review of Economic Studies.* 1994. Vol. 61. No. 3. pp. 527–544.

[136] Radner, R. "Hierarchy: The Economics of Managing." *The Journal of Economic Literature.* 1992. Vol. 30. No. 3. pp. 1382–1415.

[137] Riley, J., Samuelson, W. "Optimal Auctions." *The American Economic Review.* 1981. Vol. 71. No. 3. pp. 381–392.

[138] Rosen, S. "Authority, Control, and the Distribution of Earnings." *The Bell Journal of Economics.* 1982. Vol. 13. No. 2. pp. 311–323.

[139] Sah, R. K., Stiglitz, J. E. "The Quality of Managers in Centralized Versus Decentralized Organizations." *The Quarterly Journal of Economics.* 1991. Vol. 106. No. 1. pp. 289–295.

[140] Sah, R. K., Stiglitz, J. E. "Committees, Hierarchies and Polyarchies." *The Economic Journal.* 1988. Vol. 98. No. 391. pp. 451–470.

[141] Salanie, B. *The Economics of Contracts*, 2nd ed. Cambridge, MA: MIT Press, 2005. 224 pp.

[142] Satterthwaite, M. "Strategy-proofness and Arrow's Conditions: Existence and Correspondence Theorems for Voting Procedures and Social Welfare Functions." *Journal of Economic Theory.* 1975. Vol. 10. No. 2. pp. 187–217.

[143] Sen, A. *Collective Choice and Social Welfare*. London: Holden-Day, 1970. 254 pp.

[144] Simon, H. *The Sciences of the Artificial*, 3rd ed. Cambridge, MA: MIT Press, 1996. 247 pp.

[145] Sprumont, Y. "The Division Problem with Single-peaked Preferences: a Characterization of the Uniform Rule." *Econometrica*. 1991. Vol. 59. pp. 509–519.

[146] Stole, L. *Lectures on the Theory of Contracts and Organizations*. Chicago: University of Chicago, 1997. 104 pp.

[147] Taha, H. *Operations Research: An Introduction*, 9th ed. New York: Prentice Hall, 2011. 813 pp.

[148] *The Cambridge Handbook of Expertise and Expert Performance* (ed. by K. Ericsson). Cambridge, UK: Cambridge University Press, 2006. 918 pp.

[149] Van Zandt, T. "Decentralized Information Processing in the Theory of Organizations." In *Contemporary Economic Issues*. Vol. 4: Economic Design and Behavior, ed. by Murat Sertel. London: MacMillan Press Ltd, 1999. pp. 125–160.

[150] Vickrey, W. "Counterspeculation, Auctions and Competitive Sealed Tenders." *The Journal of Finance*. 1961. Vol. 16. No. 1. pp. 8–37.

[151] Voronin, A., Mishin, S. "Algorithms to Seek the Optimal Structure of the Organizational System." *Automation and Remote Control*. 2002. Vol. 63. No. 5. pp. 803–814.

[152] Voronin, A., Mishin, S. "A Model of Optimal Control of Structural Changes in an Organizational System." *Automation and Remote Control*. 2002. Vol. 63. No. 8. pp. 1329–1342.

[153] Wagner, H. *Principles of Operations Research*, 2nd ed. Upper Saddle River, NJ: Prentice Hall, 1975. 1039 pp.

[154] Wiener, N. *Cybernetics: or the Control and Communication in the Animal and the Machine*, 2nd ed. Cambridge, MA: The MIT Press, 1965. 212 pp.

[155] Williamson, O. "Hierarchical Control and Optimal Firm Size." *Journal of Political Economy*. 1967. Vol. 75. No. 2. pp. 123–138.

Index

A

B

D

H

I

J

K

T

U

V

W